COCONUT WATER
AND COCONUT OIL

COCONUT WATER AND COCONUT OIL

- Cook yourself healthy with coconut water, oil, milk and more
- Harness the healing powers of nature's superfood ingredient

Catherine Atkinson

LORENZ BOOKS

This edition is published by Lorenz Books,
an imprint of Anness Publishing Ltd
108 Great Russell Street, London WC1B 3NA;
info@anness.com

www.lorenzbooks.com; www.annesspublishing.com

If you like the images in this book and would like to investigate using them for publishing, promotions or advertising, please visit our website www.practicalpictures.com for more information.

© Anness Publishing Ltd 2015

All rights reserved. No part of this publication may be reproduced, stored in a retrieval system, or transmitted in any way or by any means, electronic, mechanical, photocopying, recording or otherwise, without the prior written permission of the copyright holder.

A CIP catalogue record for this book is available from the British Library.

Publisher: Joanna Lorenz
Senior Editor and Designer: Lucy Doncaster
Special Photography: Debby Lewis-Hamilton

PICTURE CREDITS
Corbis 9tl; Alamy 10t, 21tr; iStock 6bl, 6br, 7, 11, 19tl, 29tl, 30, 32, 33t, 34, 35, 38, 39t, 59b, 68t, 82t, 87t, 90b, 91t, 94t, 99b, 101t, 103t, 108b.

PUBLISHER'S NOTE
Although the advice and information in this book are believed to be accurate and true at the time of going to press, neither the author nor the publishers can accept any legal responsibility or liability for any errors or omissions that may have been made nor for any inaccuracies nor for any loss, harm or injury that comes about from following instructions or advice in this book.

SAFETY NOTE
If you have kidney problems or are on a restricted potassium intake, you should not include coconut water or other coconut products in your diet. Coconut products should never be used instead of prescribed medication.

RECIPE NOTES
• Bracketed terms are intended for American readers. For all recipes, quantities are given in both metric and imperial measures and, where appropriate, in standard cups and spoons. Follow one set of measures, but not a mixture, because they are not interchangeable.
• Standard spoon and cup measures are level.
1 tsp = 5ml, 1 tbsp = 15ml, 1 cup = 250ml/8fl oz.
Australian standard tablespoons are 20ml. Australian readers should use 3 tsp in place of 1 tbsp for measuring small quantities.
• American pints are 16fl oz/2 cups. American readers should use 20fl oz/2.5 cups in place of 1 pint when measuring liquids.
• Oven temperatures are for conventional ovens. When using a fan oven, the temperature will probably need to be reduced by about 10–20°C/20–40°F – check with your manufacturer's instruction book for guidance.
• The nutritional analysis for each recipe is calculated per portion (i.e. serving or item), unless otherwise stated. If the recipe gives a range, such as Serves 4–6, then the analysis will be for the smaller portion size, ie 6 servings. The analysis does not include optional ingredients such as salt added to taste.
• Medium (US large) eggs are used unless otherwise stated.

NUTRITIONAL INFORMATION FOR RECIPES IN THE INTRODUCTION:
Batida de Côco Energy 131kcal/548kJ; Protein 2g; Carbohydrate 9.5g, of which sugars 10.5g; Fat 3.6g, of which saturates 3.1g; Cholesterol 0mg; Calcium 52mg; Fibre 1.4g; Sodium 99mg.
Coconut and Lemon Risotto Energy 214kcal/894kJ; Protein 4.2g; Carbohydrate 41.7g, of which sugars 1.3g; Fat 3.2g, of which saturates 2.4g; Cholesterol 0mg; Calcium 32mg; Fibre 1g; Sodium 8mg.
Salmon Ceviche with Coconut Energy 159kcal/659kJ; Protein 13.5g; Carbohydrate 0.6g, of which sugars 1.7g; Fat 11.4g, of which saturates 5.1g; Cholesterol 31mg; Calcium 15mg; Fibre 1.8g; Sodium 569mg.
Coconut Pastry Energy 1253kcal/5265kJ; Protein 28.6g; Carbohydrate 174.8g, of which sugars 3.4g; Fat 53.6g, of which saturates 40.4g; Cholesterol 231mg; Calcium 349mg; Fibre 9.3g; Sodium 287mg.
Coconut Salad Dressing Energy 605kcal/2489kJ; Protein 0.8g; Carbohydrate 1.3g, of which sugars 0.5g; Fat 66.4g, of which saturates 57.1g; Cholesterol 0mg; Calcium 4mg; Fibre 0.3g; Sodium 148mg.
Coconut Mayonnaise Energy 1099kcal/4519kJ; Protein 3.1g; Carbohydrate 0.3g, of which sugars 0.3g; Fat 120.6g, of which saturates 101.1g; Cholesterol 202mg; Calcium 26mg; Fibre 0g; Sodium 83mg.
Whipped Coconut Cream Energy 88kcal/380kJ; Protein 1.2g; Carbohydrate 19.6g, of which sugars 19.6g; Fat 1.2g, of which saturates 0.8g; Cholesterol 0mg; Calcium 116mg; Fibre 0g; Sodium 440mg.
Coconut Frosting Energy 653kcal/2742kJ; Protein 0.5g; Carbohydrate 94.5g, of which sugars 94.5g; Fat 33g, of which saturates 28.6g; Cholesterol 0mg; Calcium 29mg; Fibre 0g; Sodium 16mg.
Coconut Dulce de Leche Energy 540kcal/2312kJ; Protein 1.3g; Carbohydrate 146.2g, of which sugars 146.2g; Fat 1.2g, of which saturates 0.8g; Cholesterol 0mg; Calcium 186mg; Fibre 0g; Sodium 479mg.

CONTENTS

Introducing coconut 6

Coconut water and coconut oil 8
Other coconut products 12
Using coconut water 22
Using coconut oil 24
Coconut for sport and hydration 28
Coconut for health and immunity 30
Coconut for a healthy heart 32
Coconut for weight loss 34
Coconut for beauty 36
Other uses for coconut 38

Recipes 40

Drinks and breakfasts 42

Soups 50

Snacks and salads 58

Main courses 65

Desserts 90

Baking 102

Index 112

INTRODUCING COCONUT

In the Pacific islands the coconut palm is known as the 'tree of life'. Even before anyone realized the extent of its amazing health benefits, for centuries it has been a vital natural resource for many people, both as a staple food and for trade. It is probable that Indonesia and Malaysia were the first places where coconuts grew millions of years ago. Today, coconut palm trees grow in tropical climates around the world, as far north as Hawaii and as far south as Madagascar, and their products are used almost everywhere.

Coconuts reach the rest of the world
It is likely that the coconut spread from island to island and other continents by both ship and sea. Voyagers would often take coconuts with them on their travels as, unlike most fresh food, they would keep for many months. Coconuts have a light, fibrous husk, and are both buoyant and water resistant. This allowed them to drift across oceans and be carried to distant shores. Some survived these long journeys, took root and grew in new lands.

In the 16th century, Sir Francis Drake called coconuts 'nargils', but when Spanish and Portuguese explorers first saw them they named them 'coco', meaning 'monkey face' or 'bogeyman', because the brown hairy fruit with its three indented markings looked like a head with two eyes and a mouth. The bogeyman fable was often used to scare children into good behaviour by telling them, 'si no te portas bien vendrá el coco', meaning 'if you're not good the bogeyman will come and get you'! In the mid-1700s, Samuel Johnson listed 'cocoanut' in his book *A Dictionary of the English Language*.

▼ **Below left** Coconut palms are a common sight in many tropical countries. **Below right** Coconuts derive their name from their resemblance to a monkey's face. **Bottom right** Coconut shells and oil have many uses.

INTRODUCING COCONUT

▲ Coconuts float on water, which in part accounts for their spread across the world as they drifted across seas.

This caused much confusion with cacao beans, from which chocolate is made, and in later years the 'a' was omitted from the word.

Many populations still rely on coconut for economic prosperity and on the Nicobar Islands of the Indian Ocean, coconuts were still used as currency at the beginning of the 20th century. For hundreds of years, here and elsewhere, coconuts have also been used for food, as a skin and haircare product and for their antibacterial and antimicrobial qualities, which made them suitable for medicinal purposes. During the Pacific War of 1941–5, Western doctors discovered that the benefits of coconut water included electrolyte replacement and they successfully used young coconuts as intravenous drips when medical supplies were short and sterile water unavailable. Extracts from coconuts are still included in modern medicine for a wide range of ailments.

HOW COCONUTS GROW

The coconut's name is a bit of a misnomer, as botanically it is a large fruit known as a 'drupe' and not actually a nut. Only grown successfully in sandy soil in humid, sunny climates, there are two types of coconut palm, simply known as 'tall' and 'dwarf'. Tall palm trees, which can reach over 30m/100ft in height, are grown commercially because they produce a higher number of coconuts and have a longer life, around eighty years, although it can take up to seven years before the palm produces its first coconut. Dwarf coconut palms are around a third of the size of 'tall' palms and produce coconuts after just a couple of years of planting, but are more difficult to cultivate.

The drupe grows close to the tree trunk in clusters of 10 to 12 and in most years a tree will produce at least 60 coconuts. Each takes a year to develop from flower to mature coconut; 'young' coconuts are harvested for their water content at around five months. A coconut has several layers: the smooth outer skin, which is usually green and is known as the exocarp; the next fibrous layer, which is the mesocarp; and the inner woody layer, the endocarp, which has three germination pores. When you purchase a young coconut all three layers are intact, but when buying a mature one, the first two layers have been removed, which explains why young coconuts are huge and mature ones much smaller!

COCONUT WATER AND COCONUT OIL

One of the world's most versatile foods, the simple contents of a coconut's fibrous husk are the source of an enormous range of versatile ingredients. Most are available in supermarkets and health-food stores, or can be bought by mail order. Two products are particularly renowned for their health-boosting properties: coconut water and coconut oil, but coconut in all its forms can be used instead of less-healthy fats and carbohydrates.

COCONUT WATER
The liquid in the hollow centre of a young green coconut, coconut water is not to be confused with coconut milk, which is an entirely different product. The young immature coconuts, also known as 'jelly-nuts' or 'tender nuts', are harvested when they are between five and seven months old. The opaque, almost clear liquid inside is slightly thicker than water and has a yellowish, or sometimes pinkish tinge. Subtly sweet with a hint of sourness, it has a mild nutty flavour and contains many health-giving vitamins and minerals, phytonutrients, antioxidants and enzymes, which have given coconut water its reputation and made it such a popular health-drink.

Because coconut palms usually grow near the sea, the roots have access to mineral-rich salt water as well as rainwater, so coconut water is a rich source of minerals such as potassium, calcium and magnesium and trace minerals including, iodine, zinc, manganese and selenium. One of the main benefits comes from its electrolyte content, which is why coconut water is a good sports drink. It is low-calorie, fat-free, cholesterol-free, super-hydrating, blood-pressure reducing and immune-boosting, to name just a few of its other assets.

Coconut water has long been a favoured drink throughout Latin America, the Caribbean, Asia and wherever the coconut palm grows. Now it is siphoned from its natural container and packaged in bottles, cans and cartons and exported for the rest of the world to enjoy. Some supermarkets even sell fresh young coconuts, should you want to extract unprocessed coconut water for yourself.

Coconut water is a versatile product, which can be consumed as a drink or used in cooking. Bottled and canned coconut water usually has a shelf life of at least 24 months. The flavour of brands does vary, so it may be worth trying several. Even if you buy the same brand each time, it may have a slightly different taste, as coconuts from several regions or even parts of the world may be used; season, climate and soil conditions all have an effect on the product.

Check the contents on the label carefully before buying and choose pure unadulterated coconut water; be aware that this is likely to be more expensive. Many brands are 'from concentrate', which simply means that most of

◄ Hugely versatile, coconuts provide water, oil, flour, sugar, milk and many other health-boosting products.

▲ **Top left** An Indian farmer dries coconuts to make copra, from which coconut oil can then be extracted.
Top right The freshest coconut water can be drunk straight from a young coconut.

the water has been evaporated during the production process for ease of transportation and storage; water is then added later before packing. This has little effect on the beneficial properties. Some coconut water, particularly canned products, are slightly diluted with water – usually around 15 per cent – and may contain added sugar as well. They are excellent alternatives to other canned drinks, but should be avoided when cooking savoury dishes that you wouldn't want to sweeten. Most coconut water is filtered and pasteurized and once opened it should be chilled and ideally consumed within 24 hours. For maximum benefit, drink two or three 250ml/8fl oz/1 cup servings a day, or use it in cooking.

COCONUT OIL

Solid at room temperature, coconut oil is white with a silky texture. It becomes a clear liquid when it is warmed, melting at just 24°C/75°F. The taste may be distinctly nutty, mild and aromatic or completely flavourless. More than 90 per cent of coconut oil consists of saturated fats. However, unlike the harmful long-chain fatty acids found in animal fats, the ones found in coconut oil are mainly medium-chain fatty acids, which behave differently (see page 32), and have many health benefits.

Processing can either enhance or decrease the keeping qualities, nutritional value and range of uses of coconut oil. As with other oils, all coconut oil is not the same and its quality depends on the extraction process and whether or not the oil is refined. It is important

NUTRITIONAL INFORMATION – 250ML/8FL OZ/ 1 CUP COCONUT WATER: Energy 20kcal/85kJ; Protein 5g; Carbohydrate 0g, of which sugars 15g; Fat 0g, of which saturates 0g; Cholesterol 0mg; Calcium 0mg; Fibre 7.5g; Sodium 630mg.

NUTRITIONAL INFORMATION – 30ML/2 TBSP COCONUT OIL: Energy 198kcal/813kJ; Protein 0g; Carbohydrate 0g, of which sugars 0g; Fat 22g, of which saturates 19g; Cholesterol 0mg; Calcium 0mg; Fibre 0g; Sodium 0mg.

to check labels carefully so you know exactly what you are buying. You may come across various terms including raw, cold-pressed, virgin, organic, refined and unrefined and often a combination of these.

Cold-pressed coconut oil

This is the best and most natural type of coconut oil you can buy; it has a subtle coconut taste that enhances rather than overpowers other flavours. 'Virgin' coconut oil means that the oil is generally unprocessed. You may also see it labelled as 'unrefined', which means exactly the same thing. Virgin coconut oil is made from fresh or dried coconut (known as copra), which is simply pressed to extract the oil. Although this is known as 'cold-pressing' some heat is required, but is kept below 120°C/250°F. The mixture is allowed to settle for a day or two until it separates, with the oil floating to the top. The lower temperature of extraction and minimal processing means that the coconut oil retains its antioxidants, vitamins and all its health-giving properties. 'Raw' means that the oil hasn't been heated to more than 34°C/118°F, so cold-pressed coconut oil is not necessarily raw.

▲ A traditional coconut oil mill in Kerala, India, where the oil is widely used in regional cooking.

Virgin coconut oil is one of the few types of oil that is not damaged when heated to the temperatures used in everyday cooking; it has an exceptionally high smoking point of 232°C/459°F. 'Extra-virgin' coconut oil is the same as 'virgin' coconut oil; the term 'extra' is virtually meaningless, as unlike olive oil, there are currently few industry regulations governing the production of coconut oil.

Centrifugal-extracted coconut oil

This is a simple method of extraction and usually produces high-quality coconut oil. Fresh coconut is placed in a centrifuge with hot water, then spun at high speed to separate the oil, solids and milk. Varying temperatures are used in this method, so you will need to check the label if you want to buy 'raw' coconut oil.

◀ Coconut oil is solid and white at room temperature (below 24°C/75°F), and melts to a clear oil.

Aqueous-pressed coconut oil

In this process, the coconut flesh is boiled in water until it softens and releases the oils. On cooling, the oils rise to the top. While the coconut oil is pure and doesn't contain any undesirable chemicals, the very long boiling process destroys antioxidants and other nutrients, including B-vitamins.

Expeller-extracted coconut oil

This may be good or less desirable depending on whether chemicals are used. The coconut flesh is pulverized and forced through a corkscrew press. Even though this is not a heat extraction method, the process causes a lot of friction and the temperature reaches anywhere between 60°C/140°F and 99°C/210°F. Although few of the properties of the oil are damaged at these temperatures, the oil can't be called 'raw'.

Refined coconut oil

Refining removes impurities and any chemicals that have been used during the extraction process, which is why refined coconut oil is flavourless. It is considerably cheaper than extra-virgin or virgin coconut oil and can be useful when you want to use a large amount of oil, such as for deep-frying. Refining isn't necessarily a bad thing, as modern methods include the use of steam or diatomaceous earth (a soft rock that acts as a filter) to purify oil. 'Expeller-pressed' refined coconut oil avoids the undesirable chemicals that may be used in other types of extraction.

Refining, bleaching and deodorizing (RBD) is a fairly common practice with fats and oils as many, including coconut oil, go rancid quickly (deodorizing is used to mask the unpleasant rancid smell and taste). This treatment is also used when the oil has been extracted with chemicals, which could be harmful unless they are removed. Avoid coconut oil that has been processed in this way. Unfortunately, unrefined cold-pressed virgin coconut oil and refined, bleached and deodorized coconut oil have the same milky white appearance and it is impossible to distinguish between them going by looks alone. An easy way to tell them apart is by their aroma. Virgin coconut oil smells like coconut, whereas RBD coconut oil is odourless. Unscrupulous manufacturers may of course add coconut flavouring.

BUYING COCONUT OIL

- Do not automatically assume that the label 'organic' on coconut oil means that it has not been refined; it only guarantees that the oil has not been processed using non-organic solvents and that the tree was not treated with pesticides or fertilizers according to the specific guidelines of the country in which the coconut was grown.

- Be wary if you see 'hydrogenated' or 'partially hydrogenated' on a label. This process involves combining the oils with hydrogen particles to make them more saturated, giving the oil a higher melting point and making it firmer, more stable and have a longer shelf-life, especially in hot countries. From a health point of view, hydrogenated fats should be avoided as they cause the formation of trans fats, which may contribute to heart disease and similar problems. If a product contains trans fats it should be marked on the label; they are banned in many countries.

OTHER COCONUT PRODUCTS

Available all year round, coconuts are in peak season from October to December in the southern hemisphere, where the majority are grown. The type most commonly sold in the Western world is the hard brown 'stone' of a ripe mature coconut, also known as a 'tufted coconut', but young coconuts are also available.

YOUNG COCONUT
Fresh, young, unripe coconuts are just a few months old. They have a much higher volume of coconut water inside than mature coconuts. This liquid is surrounded by a soft translucent jelly, which develops into hard coconut flesh as it matures. Young coconuts are usually green, but may also be yellow or orange. They are often sold semi-prepared with the skin cut away and resembling a white stumpy rocket, wrapped in film to retain moisture. A young coconut should be eaten within 2–3 days. They can be prepared (very carefully) in the same way as mature brown coconuts, or as below.

Preparing a young coconut
1 Place the coconut on a solid board on its side and, using a cleaver, very carefully cut a 2.5cm/1in slice off the top, keeping your hands clear of the blade. Stand the coconut upright.

2 Use the corner point of the blade nearest to the handle to make a deep incision towards the edge of the top, ensuring it goes right through.

3 Turn the coconut around 90 degrees and make another deep cut in the same way. Do this two more times, each at 90 degrees to the other two, so that you have scored a square lidded opening. Take great care doing this.

4 Using a table knife (for safety), prise out the cut square in order to open up the top. You can then drink the coconut water inside with a long straw (add a couple of ice cubes if you like) or you can pour the coconut water into a large jug, pitcher or glass.

5 Use a long spoon to scoop out the gelatinous flesh. It has a soft texture and delicate flavour and can be eaten as a healthy dessert or snack.

MATURE COCONUT
The brown shell of a tufted coconut has three eye-like germination pores at one end. Inside is a further thin brown coat called the testa and attached to this is the creamy, firm coconut flesh, also known as coconut 'meat'. The middle is hollow and filled with coconut water, which is a good guide to the freshness of the coconut; when shaken you should hear plenty of liquid inside and the coconut should feel heavy for its size. Avoid any that look, feel or smell slightly damp, especially around the 'eyes'. Stored in a cool dry place or in the refrigerator, a coconut should keep for a month, but it is better eaten within a week of purchase.

 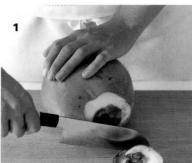 

Preparing a mature coconut

Opening a mature coconut can be tricky, so take great care when breaking into the shell to avoid injury and make sure you have plenty of space.

1 Strip off some of the hairy husk from the mature brown coconut if necessary, then put the whole coconut in a preheated oven at 180°C/350°F/Gas 4 for 15 minutes. This step isn't essential, but it will help shrink the flesh inside away from the shell, making it much easier to remove. Remove from the oven and leave until cool enough to handle.

2 To extract the coconut water, put the coconut in a small bowl, which will firmly hold it upright. Use a drill (clean the drill piece first), screwdriver or knife to pierce holes in two of the 'eyes' on the top of the coconut. Take extra care if you are using a screwdriver or knife that it doesn't slip.

3 Turn the pierced coconut upside down and allow the coconut water to drain out into the bowl positioned below. You may need to pour the collected coconut water through a fine sieve (strainer) later to remove any bits of husk or shell that may have fallen into the bowl.

4 Nestle the coconut in a thick towel to hold it steady, and firmly strike around its circumference with the back (not the blade) of a heavy knife or cleaver until the coconut cracks open. Alternatively, place the coconut in a strong, clean, plastic bag, take it outside and hit it with a hammer around the circumference.

5 Use a blunt, strong cutlery knife to prise the flesh from the shell, gouging in a direction away from your hand. You can then use a vegetable peeler to remove the brown skin from the white flesh. If necessary, rinse the flesh under cold water to remove any hairs or bits from the coconut shell.

SHREDDED, FLAKED AND DESICCATED COCONUT
All of these ingredients are produced from the white part of a mature coconut. If you prepare them yourself, they can be left fresh or be dried or toasted. Fresh coconut is delicious used in curries, salads and salsas or sprinkled over cold desserts. Dried coconut is often used in baking and confectionery and, if you don't want to make your own, can be bought unsweetened or sweetened. The long strands of shredded coconut are produced on a medium grater. Desiccated (dry unsweetened shredded) coconut is more finely grated, sometimes almost powdery. When made commercially, it is dried in hot air at around 50°C/120°F, until it contains just 3 per cent moisture. Store-bought coconut sometimes contains additives to keep the pieces separate and to retain the pure white colour; untreated coconut often has a slightly creamy tinge.

Making shredded, flaked and desiccated coconut

1 Drop the pieces of fresh coconut into a bowl of coconut water or cold water once you have removed them from the shell. This will help keep them moist and pliable until you are ready to use them. You can either peel off the brown skin or leave it on. Drain in a strainer.

2 For shredded coconut, grate the fresh coconut on a medium grater, keeping the strands as long as possible. Use a fine grater, if you want to make desiccated coconut.

3 For flakes of coconut (coconut chips), pare the fresh coconut with a peeler into long thin strips. For a more natural appearance, you can leave the brown skin on the coconut pieces rather than removing it before flaking them.

4 To toast the coconut, spread out the shredded or flaked coconut on a baking sheet lined with baking parchment and bake at 150°C/300°F/Gas 2 for 12–15 minutes, turning every few minutes so that the coconut colours evenly, until golden brown. Keep a close eye on it as coconut burns easily.

5 For sweetened coconut flakes, toss in 30ml/ 2 tbsp warmed clear honey before spreading out on the baking parchment and baking.

6 For desiccated coconut, you need to dry rather than toast the shredded coconut. Spread it out in a very thin layer on a baking sheet lined with baking parchment and bake at 120°C/250°F/Gas ½ for 12–15 minutes, stirring halfway through the drying time. The coconut will feel dry and slightly brittle when it is ready.

COCONUT BUTTER
Although coconut oil and coconut butter look similar, they are two entirely different products. Confusingly, some manufacturers use the terms interchangeably because coconut oil has a buttery consistency, so make sure you check which product you are actually buying. At room temperature coconut oil is a soft light-textured

▲ **Top left** It is easy to prepare shredded and desiccated (dry unsweetened shredded) coconut at home. **Top right** Toasting coconut flakes enhances the flavour. **Above right** Coconut butter looks similar to solid coconut oil.

white-coloured fat, whereas coconut butter is much harder. Coconut butter is made by crushing and blending finely chopped copra (dried mature coconut flesh) to make a creamy paste. It can be used as a spread and in baking, but should not be used for frying as it burns at a very low temperature. Coconut butter should not be refrigerated and will keep for up to two years. Sometimes the oil may separate from the solids and rise to the top. If this happens, give it a good stir before using it.

Making coconut butter

1 Process 250g/9oz/3 cups unsweetened full-fat desiccated coconut in a food processor for 3–4 minutes, then scrape down the sides.

2 Blend again, stopping after several minutes and scraping down the sides again. Keep repeating until the coconut oil has been extracted and the mixture has become a fairly smooth paste (it may not be quite as smooth as commercial coconut butter). This can take up to or even more than 20 minutes.

3 Transfer the coconut butter to a clean sterilized jar and store at room temperature. It will keep for at least a year.

▲ Coconut milk appears in countless recipes around the world and is sold in cans or as a powder.

▲ Coconut cream, made from the flesh, is very rich. You can buy it in cans or skim it off a can of coconut milk.

COCONUT MILK AND COCONUT CREAM

A popular ingredient in many cuisines – including Indian, Indonesian, Thai, Caribbean and Brazilian – coconut milk is not the liquid from inside the coconut, but an extraction made from the freshly grated flesh. It has a rich, creamy taste and a colour and appearance similar to that of cow's milk due to its naturally high fat content. It is made by soaking coconut with hot water, then cooling and squeezing the resulting liquid through a fine strainer to remove the solids.

In tropical countries you can buy two types of coconut milk: 'thick' coconut milk, which is made by directly squeezing grated coconut flesh through muslin (cheesecloth); and 'thin' coconut milk, where the already-squeezed flesh is then soaked in hot water for a second and third time before being strained.

In Western countries, coconut milk is usually bought in cans and is a blend of thick and thin coconut milk. Every brand is slightly different and the fat content can vary between 7.5–15ml/1½ tsp–2 tbsp per 100ml/3½fl oz. Many contain additives such as thickeners, stabilizers and emulsifiers. In the can, the coconut cream (the thick white part) and the more watery milk tend to separate, so shake the can well to blend the two before opening. You can also buy cans of reduced-fat coconut milk, which usually contain 30 per cent less fat than their full-fat counterparts.

Coconut milk can also be bought in powdered form, which can be reconstituted to the desired thickness with water, and this is particularly useful if you only require a small amount of coconut milk. It is typically mixed with double the amount of water to powder, whisking as the water is added, but can be made with less water if you want to enrich a dish with coconut milk without diluting it.

Coconut cream is the thick creamy part of coconut milk that rises to the top, in the same way that cream comes to the top of cow's milk.

OTHER COCONUT PRODUCTS

▲ Creamed coconut is a convenient and versatile way to make coconut milk and is sold as a block or in sachets.

It contains much more fat and less water. If you don't shake the can of coconut milk, you can spoon the cream from the top; this is easier if you chill the can first. You can also buy cans that contain just coconut cream. It can be used as a substitute for dairy cream in many recipes and does not separate when boiled. It has roughly the same fat content (35 per cent) as whipping cream.

Any unused coconut milk or cream should be transferred to an airtight container, stored in the refrigerator and used within 3 days.

Making coconut milk and cream

1 Remove the flesh from a mature coconut and peel away the brown skin. Chop the white flesh into pieces and place in a food processor. Pour over 150ml/¼ pint/⅔ cup near-boiling water. Blend until fairly smooth, then leave to cool for 5 minutes and blend again for just a few seconds.

2 Pour the mixture into a sieve (strainer) lined with muslin or a clean dish towel over a glass, china or plastic bowl (coconut reacts with metal) and leave it to drain. Bring the corners of the muslin or dish towel together, then squeeze to extract the milk.

3 Set aside the liquid for about 30 minutes. The coconut cream will float to the top of the milk and can be spooned off the surface. This process of soaking and squeezing can be repeated to make more (slightly thinner) coconut milk. Don't discard the coconut pulp; it can be used to make coconut flour (see page 18). One coconut will yield about 250ml/9fl oz/ 1 cup coconut milk and cream.

CREAMED COCONUT

Sold as a small solid block, creamed coconut is the ground, unsweetened, dehydrated flesh of a mature coconut that has been blended to a creamy-textured paste then compressed. Not to be confused with coconut cream, which is a concentrated liquid, creamed coconut has an intense flavour and can be stored at room temperature. It is an ideal way to thicken and add texture to curries, soups and sauces. To use, it can be grated or chopped and will melt in the heat of a dish. Stirred into hot water it can be made into coconut milk, but not cream. Sachets of creamed coconut are also available, typically containing 50g/2oz. To prepare, immerse the unopened sachet in very hot water until the creamed coconut is melted.

COCONUT FLOUR

Made from ground coconut flesh, coconut flour is gluten-free, hypoallergenic and contains almost twice as much fibre as whole-wheat flour. As it is low in digestible carbohydrates, it has a small impact on blood sugar levels, so is ideal for those watching their carbohydrate intake, diabetics and pre-diabetics, and it is also suitable for coeliacs. It is a good source of protein, containing about 19g/¾oz protein per 100g/3½oz, as well as lauric acid – a saturated fat that helps support the immune system. It also contains manganese, which helps utilize other nutrients including thiamin and vitamin C, supports bone and thyroid health and helps maintain optimal blood sugar levels.

Coconut flour is a by-product of coconut milk and is made from the leftover flesh, which is dried at a low temperature then ground to a fine flour-like powder. It has a distinctive coconut flavour and, unlike some gluten-free flours, is not easy to incorporate into regular recipes without making other adaptations. It absorbs a huge amount of liquid, so when baking you will probably need to add at least the equivalent amount of liquid to the amount of coconut flour. Because it is gluten-free, you will also need to use extra binding ingredients such as eggs, or sweeteners such as honey or maple syrup. Either use recipes that have been specifically developed using coconut flour or, if you want to use it in your regular baking recipes, start by substituting 10–30 per cent of other grain-based flours with coconut flour and add more egg and liquid. It can also be used as a thickening agent for sauces and gravies. It has a long-shelf life and will keep for up to a year if stored in a cool, dark, dry place.

Making coconut flour

1 Preheat the oven to 110°C/225°F/Gas ¼. Line a baking sheet with baking parchment (don't use foil as the metal may taint the flavour of the coconut). Spread the leftover pulp from making coconut milk (see page 17) over the baking parchment in a thin, even layer, using a fork to break up any lumps.

2 Bake for 15 minutes, then remove from the oven and stir with a fork. Spread out again into an even layer and bake for a further 15–20 minutes, or until it is completely dry. Coconut can burn even at a low temperature, so check it often towards the end of the time.

3 Remove the coconut from the oven and leave until completely cool. Tip into a food processor and process for 4–5 minutes or until it is very finely ground. Store the flour in an airtight container and use within one year.

COCONUT SUGAR

Sometimes labelled 'coconut palm sugar' or 'crystallized coconut nectar', coconut sugar is subtly sweet and similar to brown sugar in appearance and taste but with a hint of caramel. It has been used as a natural sweetener for thousands of years in countries where the

▲ **Top left** Coconut sugar is made from the sap of the coconut tree's blossoms and flower buds. **Top right** Coconut flour. **Above right** Coconut sugar.

coconut palm grows. Because it is not very processed, the colour, flavour and sweetness can vary between batches and brands, as much depends on the coconut species and the conditions in which is has been grown. It comes as sugar crystals, or in block or liquid form.

Coconut sugar is produced from the sap of flower buds and blossoms of the coconut tree and is still collected by hand, as it has been for decades. A cut is made on the spadix (the spike of the fleshy stem) and the sap is collected in bamboo containers. This is then boiled to evaporate most of the moisture until it becomes a thick syrup known as 'toddy'. From this it can be further reduced to a soft paste, harder block or crystal form or left as a syrup, which you will find labelled 'coconut syrup' or 'coconut blossom syrup'. This is similar in taste and texture to maple syrup but is a slightly darker colour.

Coconut sugar has a Glycemic Index (GI) of 35 – compared to refined white and brown cane sugar, which has a GI of 68 – so is a healthier option, especially for diabetics as it will not spike blood-sugar levels in the same way as other sugars. Use as a substitute for sweetening drinks, and in cooking and baking.

Coconut sugar has a high mineral content and is especially rich in iron, zinc, potassium and magnesium. When compared to brown cane sugar, it has 36 times more iron and 10 times more zinc. It also contains a number of B vitamins – notably B1, B2, B3 and B6 – although some of these will be destroyed during the boiling process. Coconut sugar contains 16 of the 20 amino acids, although some are in trace amounts, and it is especially high in glutamine, which is important for metabolism. In some countries the term 'coconut sugar' and 'palm sugar' are used interchangeably, but the two are different, so check you are buying genuine coconut sugar; the label should state that it is 100 per cent coconut sap or coconut palm.

COCONUT-FLAVOURED MILKS AND DRINKS

An increasing number of coconut-based drinks is appearing in supermarkets. These include ones that are intended as a non-dairy alternative to milk to be poured over breakfast cereals, used in milkshakes, smoothies and in cooking, and as a drink in their own right. Most are a blend of water, coconut milk and rice milk. Generally, they are low in calories – around 20kcal/85kJ per 100ml/3½fl oz – and contain little or no added sugar. Many are fortified with vitamins and calcium.

Blends of fruit juice and coconut, and coconut smoothies, are also popular. Some contain pure unsweetened fruit juice combined with coconut water or coconut milk; others are fruit-flavoured drinks with coconut flavouring and added sugar. Check the labels carefully.

▼ **Bottom left** Coconut drinks are available in many different flavours. **Bottom right** Coconut aminos and coconut vinegar can be used instead of soy sauce and other types of vinegar respectively.

NATA DE COCO

This chewy translucent jelly is usually sold cut into small cubes. Nata de coco is made by fermenting coconut water, during which process the gel is formed through the natural production of microbial cellulose. Originating in the Philippines, its Spanish name means 'cream of coconut'. It is often added to canned fruit such as pineapple, where it imparts a subtle coconut taste and extra texture.

COCONUT VINEGAR

This is a staple flavouring in South-east Asia, particularly the Philippines where it is called 'suka ng nyog'. It is made from either sap collected from coconut flowers or fermented coconut water, which is blended with sugar and then pasteurized. After the addition of yeast it is fermented for a week, then a vinegar starter culture is added and it is fermented for a further week. This produces a white-coloured, cloudy, sweet vinegar with low acidity. The best coconut vinegars contain live cultures.

COCONUT AMINOS
Made by fermenting the sap from coconut flowers, coconut aminos is a salty flavouring that can be used as a substitute for soy sauce. High in amino acids, it is wheat- and soy-free, so it is suitable for those on paleo diets. Once opened, store in the refrigerator.

COCONUT ALCOHOL
Coconut drinks and liqueurs became popular during the cocktail renaissance of the 1980s. One of the best-known is Malibu, made in Jamaica, which is a blend of Barbados white rum and coconut for which the dried pulp and milk of the coconut are used. Another is Batida de Côco, a cocktail made in Brazil.

Coconut sap or nectar taken from the clusters of coconut flowers can be fermented to make coconut wine. This may be distilled to make a spirit known as 'arrack' (not be confused with 'arak' which is an anise-flavoured alcoholic drink). Maturing the wine in barrels gives arrack a rich golden colour and mellow flavour.

▲ **Top left** Arrack can be served simply as shots. **Above left** Nata de coco has a chewy texture. **Top right** A traditional still for making arrack from the sap of the Palmyra or 'toddy palm', near Bagan, Burma.

BATIDA DE CÔCO
To make this Brazilian cocktail using arrack instead of the traditional cachaça, blend 400ml/14fl oz/1⅔ cups coconut milk, 150ml/¼ pint/⅔ cup condensed milk, 200ml/7fl oz/scant 1 cup coconut water, 115g/4oz/1 cup freshly grated coconut and 300ml/½ pint/1¼ cups arrack until smooth. Transfer to a glass bottle, ensuring there is a gap of 5cm/2in between the liquid and top of the bottle. Freeze for 1 hour before serving icy cold. It serves 12 people.

USING COCONUT WATER

The simplest way to enjoy coconut water is as a drink, but there are lots of other ways it can be used. If you are not a fan of the beverage, you may still enjoy all the benefits, as coconut water can easily be incorporated into everyday eating: in delicate dishes, the mild, slightly sweet and nutty flavour of coconut water can enhance the other ingredients; in more highly flavoured food it will be barely noticeable.

The recommended daily amount is 500ml/17fl oz/generous 2 cups–750ml/1¼ pints/3 cups; two to three medium glassfuls. Consuming a smaller amount will still be beneficial, but do not drink more – coconut water should be part of a well-balanced, healthy-eating plan. If you have kidney problems or are on a restricted potassium intake, do not include coconut water or other coconut products in your diet.

▼ Coconut water makes a refreshing, nutritious and hydrating drink and is best served ice-cold.

EASY WAYS TO ADD COCONUT WATER TO YOUR DIET

- Substitute up to half of the fruit juice in a fruit smoothie with coconut water.

- Coconut water can be served as a lactose-free alternative to cow's milk on cereals or for making porridge. Its sweet flavour reduces the need for additional sugar. It doesn't curdle when mixed with milk, so you can use a combination of dairy or soya milk and coconut water.

- Replace up to half of the stock or other liquid in home-made soup with coconut water. It goes especially well with vegetable soups such as corn, sweet potato, butternut squash and tomato and is good with chicken and seafood.

- Use coconut water for part or all of the liquid if using the absorption method to cook rice. It can be used in risottos, but consider flavour combinations carefully.

- Many curry-type recipes contain full-fat coconut milk. Swapping up to half of this for coconut water will still give you all of the flavour but less fat.

- Soak dried fruits, such as dried apricots, prunes, apples, pineapple and mango in coconut water to make a delicious high-fibre dessert.

- Use small amounts of coconut water when making custard-based desserts or frozen ones such as ice cream and sorbet.

- Try using coconut water instead of some of the milk in muffins and batter mixtures, such as for pancakes.

Coconut and lemon risotto

Coconut water adds a subtle nutty flavour to this risotto and its slight sweetness helps balance the richness of the dish.

475ml/16fl oz/2 cups coconut water
450ml/¾ pint/scant 2 cups vegetable stock
15ml/1 tbsp coconut oil
1 small onion, finely chopped
1 clove garlic, crushed
1 stick celery, finely chopped
200g/7oz/1 cup risotto rice
finely grated rind and juice of ½ lemon
45ml/3 tbsp chopped fresh herbs, such as parsley or coriander (cilantro)
salt and freshly ground black pepper

Serves 3–4

1 Pour the coconut water and stock into a pan and heat until steaming. Turn down the heat as low as possible to keep it hot; do not boil.

2 Heat the oil in a large pan, then add the onion, garlic and celery and cook gently for 4–5 minutes or until softened, stirring from time to time. Add the rice and cook for 1 minute, stirring to coat the grains with the oil.

3 Add a ladleful of the liquid to the pan and bubble until it has almost all been absorbed, stirring often. Continue in the same way for 15–20 minutes, until the rice is just tender. Stir in the lemon rind and juice, herbs and seasoning.

Salmon ceviche with coconut

In this delicious dish, the salmon is 'cooked' by the acid in the lime juice and coconut.

250g/9oz very fresh salmon fillet, skinned
30ml/2 tbsp rock salt, and ground black pepper
1 red chilli, seeded and finely chopped
squeezed juice of 4 limes
75ml/5 tbsp coconut water
50g/2oz freshly shredded coconut (see page 14)

Serves 4

1 Sprinkle the salmon with salt and chill in the refrigerator for 20 minutes. Wipe off the salt, then rinse the fish and pat dry. Slice thinly at an angle.

2 Arrange the salmon slices on a serving plate and sprinkle over the chopped chilli. Season to taste with freshly ground black pepper.

3 Whisk together the lime juice and coconut water, then use to coat the fish. Chill for 30 minutes, until the fish is opaque. Top with shredded coconut and serve immediately.

USING COCONUT OIL

Coconut oil is versatile and for the most part it can be substituted in equal parts for any other fat or oil. Since it has a low melting temperature, around 24°C/75°F, it changes from solid to liquid and back again easily. You can place the entire jar in warm water when you want a few tablespoons then let the rest re-solidify.

When using in liquid form, make sure that chilled ingredients such as eggs and milk are at room temperature before mixing, or they may cause the oil to re-solidify. If you need to melt coconut oil, make sure other liquids are at a similar temperature; warming them together is often easiest. Coconut oil doesn't re-set quite as hard as solid fats such as butter or baking margarine, and some bakes, ie those with crumbly cookie bases, may be slightly soft; storing in the refrigerator will firm them.

Although raw virgin coconut oil has the greatest health benefits, many modern refining methods have little negative impact on the oil, so it's worth stocking both types. The latter is flavourless and is preferable for recipes for which you don't want to add a coconut flavour.

To enjoy the maximum health benefits of coconut oil, the recommended daily amount is between 30ml/2 tbsp and 60ml/4 tbsp depending on your body weight. Eating a higher amount will not be more beneficial.

▼ Coconut oil has a low melting temperature and changes from a white solid to a clear liquid.

EASY WAYS TO ADD COCONUT OIL TO YOUR DIET

- Try virgin coconut oil as an alternative spread to butter or margarine.

- Brush coconut oil over food before grilling (broiling) or roasting. You can warm it a little first to make this easier, adding fresh or dried herbs, spices or grated citrus zest for extra flavour. Tuck a little under the breast skin of whole chicken to baste it during cooking.

- Use coconut oil for roasting potatoes and other root vegetables. Use a good-quality refined coconut oil if you don't want them to have a coconut flavour.

- To make a marinade for meat, chicken or fish (marinate fish for no more than 30 minutes), gently heat coconut oil and coconut vinegar using twice as much oil to vinegar, until the oil has melted. Add other flavourings if liked. Leave until just cool, then combine with the meat, poultry or fish.

- Coconut oil makes a fantastically nutty pesto sauce and is also good for dips such as hummus; simply replace melted coconut oil for all or part of the olive oil.

- For a sweet topping for home-made popcorn, gently heat together equal quantities of coconut oil and coconut sugar. Drizzle over the popped corn and mix until evenly coated.

- Coconut oil is a superb substitute for butter when brushing filo pastry as, unlike butter, it doesn't contain any liquid so the baked results will be much crisper.

Coconut pastry

Coconut oil makes a crisp pastry. It's very dry when you first make it, then gets moister when covered with a damp dish towel. Make sure that the egg is at room temperature or it may cause the coconut oil to solidify. This quantity is enough to line the base and sides of a 23cm/9in flan tin (quiche pan).

225g/8oz/2 cups plain (all-purpose) flour
a pinch of salt
60ml/4 tbsp coconut oil
30ml/2 tbsp warm water
1 egg, at room temperature

Makes about 275g/10oz

1 Sift the flour and salt into a bowl. Put the coconut oil in a small bowl and place in a pan with warm water coming halfway up the sides of the bowl. When the coconut oil has melted to a clear liquid, remove from the pan, add the water and egg and whisk together with a fork.

2 Make a well in the centre of the flour, add the egg mixture and stir with a fork, gradually incorporating the flour until all the liquid has been worked in and a dough is formed.

3 Knead on a lightly floured surface for a few seconds until smooth, then cover with a damp dish towel and leave to rest for 30 minutes before using. Knead only until smooth and do not overwork, or the pastry will be tough.

Coconut salad dressing

This creamy oil and vinegar dressing can be drizzled over leafy salads or used as a marinade for lamb or chicken.

90ml/6 tbsp coconut oil
5ml/1 tsp Dijon mustard
1 small clove garlic, crushed
45ml/3 tbsp coconut vinegar or balsamic vinegar
salt and freshly ground black pepper

Makes about 150ml/¼ pint/⅔ cup

1 Melt the coconut oil by placing it in a glass jar, then put the jar in a pan with warm (not boiling) water coming halfway up the sides of the jar.

2 Put the mustard and garlic in a bowl. Gradually whisk in 15ml/1 tbsp of the vinegar. Whisk in the oil in a steady stream, then the remaining vinegar. Season.

3 Alternatively, put all the ingredients in a screw-topped jar and shake until thoroughly blended. Store the dressing at room temperature. Use within 2 days of making.

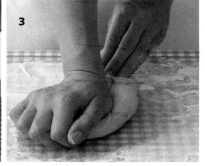

Coconut mayonnaise

Coconut oil makes a rich, creamy mayonnaise. Make sure all the ingredients are at room temperature before you start and add the oil drop by drop, or the mayonnaise may curdle.

150ml/¼ pint/⅔ cup coconut oil
1 egg yolk
5ml/1 tsp lemon juice
2.5ml/½ tsp Dijon mustard
salt and pepper

Makes about 150ml/¼ pint/⅔ cup

1 Melt the coconut oil by placing it in a glass jar and putting it in a pan with warm water coming halfway up the sides of the jar.

2 Put the egg yolk, lemon juice, mustard, salt and pepper in a small bowl. Whisk together.

3 Start adding the oil, one drop at a time, whisking until about a third has been added and the mixture begins to thicken, then pour in the rest of the oil in a thin, slow, steady stream, whisking constantly until thick and glossy. Store covered in the refrigerator for up to 5 days.

Cook's tip
You can also make this mayonnaise in a food processor. Place all the ingredients except the oil in the processor and blend briefly until pale and creamy. With the motor running, slowly add the melted oil in a thin stream until blended.

Whipped coconut cream

This low-fat alternative to dairy whipped cream is very rich, so serve in small helpings. This method does not work as well with canned coconut cream, so use the top of coconut milk, as outlined below.

400ml/14oz can full-fat coconut milk, chilled overnight
5ml/1 tsp vanilla extract

Makes about 150ml/¼ pint/⅔ cup

1 Carefully open the can; the coconut cream will be at the top. Lift it out and place in a chilled glass bowl, keeping the liquid for other recipes.

2 Add the vanilla extract to the bowl, then whisk for a few minutes with an electric beater until the mixture is light and soft peaks form.

3 Serve the whipped coconut cream immediately, perhaps spooned over fruit, or cover with clear film (plastic wrap) and store in the refrigerator for 2–3 days.

Coconut frosting

This is a dairy-free alternative to less-healthy buttercream and can be spread over cakes and cookies or piped. On a hot day, or if the coconut oil is very soft, chill it first for about an hour.

50g/2oz/¼ cup virgin coconut oil
90g/3½oz/scant 1 cup icing (confectioners') sugar, sifted
5ml/1 tsp vanilla or coconut extract
10–15ml/2–3 tsp coconut milk or soya milk
2–3 drops natural food colouring (optional)

Makes enough to frost 12 cupcakes

1 Put the coconut oil in a bowl and beat for a few seconds with a wooden spoon. Sift over about a third of the icing sugar, add the vanilla or coconut extract and stir until mixed.

2 Gradually stir in the rest of the icing sugar along with 5ml/1 tsp coconut or soya milk. Beat well, and add coconut or soya milk if needed. This depends on whether you are planning to spread or pipe the frosting.

3 If liked, add 2–3 drops of food colouring. Beat until the frosting is light and completely blended. Spread or pipe as needed. Store any unused frosting in the refrigerator for 2–3 days.

Variation
For chocolate frosting, blend 20ml/4 tsp boiling water with 10ml/2 tsp (unsweetened) cocoa powder. Cool, then beat into the frosting.

Coconut dulce de leche

This is quite calorific, so should be eaten in moderation as a special treat. It is fantastic combined with bananas in a banoffee pie or serve as a dip with fruit.

400ml/14oz can full-fat coconut milk
75g/3oz/⅓ cup coconut sugar
50g/2oz/¼ cup caster (superfine) sugar
a pinch of salt

Makes just over 150ml/¼ pint/⅔ cup

1 Put the coconut milk, sugars and salt in a medium heavy pan. Gently heat, stirring now and then until the sugar has completely dissolved.

2 Bring to a boil, reduce the heat to low and simmer, stirring occasionally for 20–25 minutes, until it darkens to a golden-caramel and thickens.

3 Remove the pan from the heat and allow the mixture to cool, stirring occasionally; it will thicken further as it cools. Cover and store in the refrigerator for up to a week.

COCONUT FOR SPORT AND HYDRATION

Although coconut water has been around for millennia, it is only in the last few years that its attributes have been recognized and sales around the world have rocketed. It is not only deliciously refreshing, but is beneficial post-exercise because it contains a lot of natural electrolytes, meaning it is a fast-hydrating drink.

Electrolytes are inorganic compounds that become ions in solution and are important for sending electrical impulses between different parts of the brain, nervous system and heart. The balance of water and your blood pH (degree of acidity or alkalinity) depends on electrolytes. When you exercise heavily you lose electrolytes, especially potassium and sodium, in your sweat and these need to be replenished. Although drinking water can hydrate you, it doesn't have the speed of hydration of coconut water because it doesn't have electrolytes in it.

COCONUT WATER CONTAINS FIVE ELECTROLYTES:
1. Potassium This is the electrolyte that coconut water is renowned for. 250ml/8fl oz/ 1 cup contains as much potassium as two bananas, which are another natural source that contains high levels. Potassium can maintain water pressure within the cells and blood by helping draw water into both; it enables the body to rehydrate quickly and reduces the feeling of fatigue and muscle-ache after exercise. It can also help prevent cramp and spasms in the muscles. Higher intakes of potassium may counteract the effects of sodium and therefore reduce high blood pressure. It is almost impossible to consume too much potassium as any excess is excreted. RNI: 3.5g

2. Sodium Vital for controlling the amount of water in the body, maintaining the normal pH of blood, transmitting nerve signals and helping muscular contraction, sodium is present in all foods in various degrees and is a component of table salt. Coconut water has less sodium than is found in most sports drinks, which is good news for those watching their salt intake (most of us consume too much). However, if you are an intensive athlete, and wish to increase the amount of sodium a little, use a combination of coconut water and a sports drinks. RNI: 1600mg

3. Calcium This is the main mineral present in bones and teeth. Studies have shown that when the dietary intake of calcium is increased, there is also an increase in the breakdown of fat and a decrease in the production of fat. This is good news if you are exercising for weight loss. RNI: 700mg

4. Magnesium This is concentrated mainly in the bones and muscles, but is also an essential part of all cells. It's needed for energy release,

◀ Your body loses electrolytes through sweating while exercising, and these need to be replenished.

▲ **Top left** Coconut water is a great drink during hot weather as well as after exercise. **Top right** A glassful of coconut water is packed with electrolytes.

cell division, enzyme production and nerve and muscle function. Studies suggest that magnesium can help lower blood pressure as well. In addition, magnesium helps the body process fat and protein and makes protein, which is especially important if you are trying to build and maintain muscles. RNI: 270mg (women), 300mg (men)

5. Phosphorus After calcium, this is the most abundant mineral in the body. It is present in all cells and has a fundamental role in converting the energy from food to a form that can be used by the body. RNI: 540mg

RNI (Reference Nutrient Intake) This is the amount of a nutrient that is enough for almost everybody, including people with higher needs. 250ml/8fl oz/1 cup of coconut water contains 50kcal/209kJ, 665mg of potassium, 275mg of sodium, 65mg calcium, 65mg of magnesium, 50mg of phosphorus and 2g of protein.

NATURAL HYDRATION
Coconut water has gained popularity as an alternative to sports drinks because it is a natural product, low in sugar and free of artificial colours and caffeine. Its electrolyte content is more than double that of traditional sports drinks with about half of the carbohydrates. If you are exercising in order to lose weight, remember that although coconut water has relatively no fat content and is much lower in sugar than other fruit juices (it has less than a fifth of the sugar content of other juice), it is not calorie-free, so drink plain water too.

Even if you are more of a couch potato than a sports fanatic, you can still benefit from coconut water as the potassium content can help counteract the problems associated with a high-sodium diet of processed foods.

FRACTIONATED COCONUT OIL
Sometimes referred to as liquid coconut oil, fractionated coconut oil is a product derived from coconut oil. The medium-chain triglycerides (MCT) – fatty acids – are separated out of the coconut oil and sold as a different product. Fractionated coconut oil is promoted as a product for 'athletes' as a rapid source of energy for muscle gain for those on a low-carbohydrate diet. Do check the label or online information carefully and make sure the product you are buying is suitable for consumption, as fractionated coconut oil is also used as a carrier oil in beauty products. If you are using it for fitness purposes, you will find it in health-food and sports stores.

COCONUT FOR HEALTH AND IMMUNITY

Coconuts have antibacterial, antiviral and antifungal properties that can help destroy harmful bacteria, viruses and fungal infections. Coconut oil can also boost your immune system, helping prevent illnesses. More than 90 per cent of coconut oil is saturated fat and around two-thirds of that comes from medium-chain triglycerides (MCTs). The main component of these MCTs is lauric acid, which the body converts into monolaurin. This is helpful for dealing with viruses, bacteria and fungi and yeast that cause infections and diseases such as colds, influenza and candida, and may help fight herpes and even slow the progress of HIV and cancer. In addition, it may help to speed recovery if you are already unwell. Apart from coconut, the only other place lauric acid is found is breastmilk.

There are other MCTs in coconut: capric acid and caprylic acid. Converted to monocaprin and monocaprylin they also hinder microbes from attacking the immune system. Individually these three MCTs attack microbes differently, but combined they will provide your body with a protective shield. It is important to realize that consuming a single food will not give you a strong immune system, but when added to a healthy diet and lifestyle, it can boost your immunity. NOTE: coconuts should never be used instead of prescribed medication.

FIGHTING GERMS AND BACTERIA

Many illnesses and diseases are caused by bacteria, both of which spread by multiplication, and include ulcers, throat and urinary tract infections and pneumonia. Monolaurin (a monoglyceride produced from lauric acid in coconut) packs a strong antibacterial punch. It binds to the lipid membrane that covers the virus or bacteria, slowing and even stopping its replication. Because it is similar in composition to the fatty acids in the bacteria coating, the MCTs are absorbed by the virus and the lipid membrane is weakened until it breaks apart and is destroyed.

FIGHTING VIRUSES

There are over 200 viruses that can cause a cold, and although not serious, the symptoms are unpleasant. While you cannot cure a cold, one of the best ways to lessen the severity and the time it lasts for is to eat foods that support your immune system. Eat things that are rich in antioxidants, such as colourful fruit and vegetables, together with coconut for its antiviral properties and antioxidant vitamins E and C. Vitamin C lowers histamine levels in the body, which helps reduce symptoms. You should also drink plenty of fluids, including coconut water, to remain hydrated.

FIGHTING YEAST INFECTIONS

There are many strains of yeast but only one causes health problems in humans – candida albicans. This yeast present in the gut and mouth can cause thrush; ear, nose and throat infections; inflammation; itchy, dry skin; and problems with your nails and hair. Caprylic

◀ Coconut products can help keep you healthy when consumed as part of a balanced diet.

COCONUT FOR HEALTH AND IMMUNITY

▲ Coconut oil contains medium-chain fatty acids that the body converts into the beneficial compound monolaurin.

acid, which is found naturally in coconut, has been shown to prevent and alleviate candida. If you frequently suffer from yeast infections try to reduce or exclude high-sugar and refined and processed food. Coconut oil can also be applied externally to treat ringworm, athlete's foot, nappy rash and fungal nail infection.

COCONUT FOR DIABETES

Diabetes is a condition that affects how the body uses carbohydrates for energy and turns them into glucose. For glucose in the bloodstream to be able to enter cells, the hormone insulin is needed. Sometime the body cannot make enough insulin and this causes blood glucose levels to rise. Diabetics need to control their blood sugar, blood pressure, cholesterol and triglycerides levels to avoid complications and coconut can help with all these. Diabetics should choose low-glycaemic carbohydrates, which take longer to break down in the body, reducing the risk of high blood-sugar levels. Coconut sugar is a great replacement for refined sugar as it is a low-glycaemic food with a glycemic index number of 35 (under 55 is considered a low-glycaemic food). Pair it with coconut flour for healthier baking.

OTHER HEALTH BENEFITS OF COCONUT

- Coconut can help to protect the body from harmful free radicals that can cause degenerative diseases and ageing.

- A thin layer of coconut oil applied to the skin helps reduce symptoms of skin problems such as eczema and dermatitis.

- Coconut oil can be used as a treatment to kill head lice. Apply over hair, leave for an hour, then use a nit-comb to remove the lice; they will be unable to cling to the hair and lay further eggs. Wash hair and hot-wash pillowcases and bedding. Repeat the following day.

- Coconut may help fight against cancer by protecting our immune system. In addition, coconut water contains cytokinins, which have anti-carcinogenic properties, and the nutrient kinetin riboside, which can curb the growth of multiple myelomas (skin cancer).

- Coconut may help treat HIV and AIDs. These weaken the immune system and allow secondary infections to set in, which can cause illness and death. Coconut oil provides a dietary supplement that can strengthen the immune system and may protect the body from further illness.

COCONUT FOR THE PALEO DIET

The Paleolithic, 'caveman' or 'hunter-gatherer' diet is based on the premise that we should eat only what our ancient ancestors ate. Followers believe that we are not sufficiently evolved to properly digest 'new foods' such as cultivated grains, dairy or processed foods. Coconut products are suitable for those on this diet.

COCONUT FOR A HEALTHY HEART

According to the World Health Organization, heart disease is the leading cause of death in the world. The most common reason for heart disease is the hardening and thickening of arteries from plaque build-up, known as 'arteriosclerosis'. This furring of arteries constricts blood flow and may lead to a heart attack or stroke. While over three-quarters of those affected are over the age of 65, heart attacks in the under-50s are on the increase.

Many will be amazed to hear that coconut is actually good for your heart. For decades nutritional guidelines have been very clear: we should all eat less fat, especially saturated fat. We have been told to replace saturated fats with unsaturated fats such as monounsaturated and polyunsaturated fats including those found in vegetable oils. Yet many populations, such as Hawaiians and Filipinos, rely on coconut as a staple in their diet and consume far more than the recommended maximum 10 per cent of daily calories coming from saturated fat. Indeed, many consume up to six times this amount, yet have an amazingly low rate of heart disease. How, then, can coconut be beneficial when more than 90 per cent of the fat is saturated?

▼ Coconuts contain 'good' cholesterol that can help protect the heart against disease.

Research has discovered that coconut's saturated fat is different from any other type of fat, so it affects the heart differently. Fats and oils are composed of molecules called fatty acids, which are made of chains of carbon atoms with hydrogen atoms attached. Some fatty acids are smaller than others; they may be short-chain, medium-chain or long-chain.

The size is important because our bodies metabolize each fat differently depending on its size. Most of the oils and fats that we eat, regardless of whether they are saturated or unsaturated, are long-chain fatty acids. These must be broken down into smaller lipoproteins before they can pass through the intestinal walls and be released into the bloodstream. Because of this, the body takes time to process the fatty acid into energy, so initially stores it as body fat. This in turn may add to the build-up of arterial plaque. Coconut oil has medium- and short-chain fatty acids that are digested and absorbed differently to long-chain fatty acids and are considered healthier. They are a quick source of energy and are digested in a similar way to carbohydrates rather than being stored.

THE CHOLESTEROL QUESTION

Cholesterol is a fat-like substance that plays a vital role in the body. It's the material from which many essential hormones and vitamin D are made. Cholesterol is carried around attached to proteins called high-density lipoproteins (HDL), low-density lipoproteins (LDL) and very-low-density lipoproteins (VDL). After eating, the LDLs carry the fat in the blood to the cells where it's required. Any surplus should be excreted from the body; however, if there is too much LDL some of the fat will be deposited on the walls of the arteries. In contrast, HDLs appear to protect against heart disease.

COCONUT FOR A HEALTHY HEART

▲ To keep your heart healthy, try to do moderate exercise, such as swimming, for at least 2½ hours per week.

Coconut oil contains a fat known as lauric acid, a medium-chain fatty acid, which increases the good HDL cholesterol in the blood and helps improve the HDL/LDL ratio, which is an indicator of the risk of heart disease.

COCONUT OIL STABILITY

There has been much research on free radicals and the damage they can do to healthy cells. Free radicals can arise as a result of 'oxidative stress', which is caused by external toxic sources (such as pollution) or from internal sources (such as the food we eat). Unsaturated oils are unstable and changes in storage temperatures and the heat of cooking can affect them and make them rancid. However, in the early stages it is unlikely that you would detect any changes in the flavour or colour of the oil.

Eating foods that contain rancid oils increases free radical activity that can cause inflammation and could in turn adversely affect heart health. Coconut oil, however, does not need to be chilled and is so stable that it can be stored at room temperature for several years. It is also high in vitamins, minerals and antioxidants (which help to fight free radicals), all of which are good for your heart.

TOP TIPS FOR A HEALTHY HEART

- **Eat a well-balanced diet** Use coconut oil where possible instead of other oils and butter (see pages 24–7). Make sure you have oily fish twice a week; salmon, mackerel, fresh tuna and sardines are an excellent source of omega-3 fats, which can help protect against heart disease.

- **Control your weight** Being overweight can increase your risk of heart disease, so try to keep within a healthy weight range (see pages 34–5 for how coconut oil can help you lose weight).

- **Get active** Do 2½ hours of moderate exercise every week. This can be sports such as cycling or swimming, but remember that brisk walking, vigorous housework or gardening all count too.

- **Give up salt** To maintain a healthy blood pressure, stop adding salt to food and use less when cooking. Watch for high salt levels in processed foods; a food is high in salt if it has more than 1.5g salt (or 0.6g sodium) per 100g/3¾oz.

- **Quit smoking** Smoking is a major contributor to coronary heart disease. After just one year of giving up, your risk of heart disease falls to about half that of a smoker.

COCONUT FOR WEIGHT LOSS

Combined with a healthy-eating plan, coconut can help promote weight loss. The dynamics of weight gain and loss are simple; if you consume more calories than you use, you will gain weight, if you consume less than you use, you will achieve weight loss. Ignore anyone who talks about 'eating the wrong foods' or 'eating at the wrong time of day'; a calorie is a calorie, no matter what food it is found in or when it is consumed. Of course, if you eat a lot of high-fat, high-sugar foods it is easier to consume more calories than you need without realizing, and if you spend your evening eating bags of snacks in front of the television you are likely to exceed the amount of calories you will have burnt off during the day.

Excess food intake is stored in the body as fat, leading to weight gain and eventually to obesity. This in turn, can lead to various medical conditions such as cardiovascular disease, diabetes, joint problems and cancer. Losing weight isn't easy and there are many factors that make it difficult, including a sedentary lifestyle, social eating, 'comfort' eating and cravings; these can break the strongest willpower.

Rather than trying out the latest diet fad, or crash-dieting, incorporating coconut into your diet can help you lose weight gradually and should be seen as a lifestyle change. Coconut oil contains medium-chain fatty acids (see page 32), which are easily absorbed and a preferred source of energy as long as you do not consume an excessive amount of calories at the same time. The fat can go straight to the cells and be used as food immediately, and this can help stimulate your metabolism to burn more fat. Studies have shown that those who have diets high in medium-chain fatty acids have a higher thermogenic (fat-burning) rate within the body. However, like any fat, coconut oil is calorie-dense (around 9 calories per gram), so if you are trying to lose weight, use it in moderation as an alternative and not in addition to other fats.

WHY YOU SHOULD AVOID VERY LOW-FAT DIETS
Many people are concerned that if they consume more fat it will prevent weight loss. During the 1980s and 90s very low-fat diets were all the rage. Such restrictive diets are rarely successful; sticking to them for more than a few weeks is beyond most of us, which is why, ultimately, most fail. Our bodies need and demand fat in order to remain healthy.

Fat forms a major part of all cell membranes and is vital for the absorption of the fat-soluble vitamins A, D, E and K. If you eat a very low-fat diet you may have major and minor health problems including a poor immune system, inability to sleep, depression, dry skin and low energy levels.

◀ Eating healthy meals at regular intervals will help prevent snacking. Try using coconut water in place of milk on your cereal for a great start to the day.

HOW COCONUT OIL SUPPORTS THYROID FUNCTION

Coconut oil is thought to have a balancing effect on the thyroid and it is possible that its unique medium-chain fatty acids improve the body's ability to use fats and sugars by raising metabolism and stabilizing blood sugar levels. The thyroid plays a huge role in controlling metabolism and weight loss can be difficult for those who suffer with hypothyroidism (low-thyroid function). You may have an underactive thyroid disorder if you experience unexplained weight gain, constant fatigue, hair loss and insomnia and should get this checked out by a doctor.

WHY COCONUT MAKES YOU FEEL FULL FOR LONGER

Satiety is the feeling of fullness you should get after a meal, when the hormone leptin sends a signal to your brain that you have eaten enough. Sometimes we over-rule the signal and are tempted to eat more even though we are already full. Fats and proteins are more complex than carbohydrates, so help make you feel full more quickly. Many coconut products are also high in insoluble fibre (around 93 per cent of the coconut's fibre is insoluble). Because the body is unable to digest this it takes longer for the food to move from your stomach to your intestines, so the feeling of satiety lasts much longer and helps to reduce food cravings. Using products like high-protein, high-fibre coconut flour will allow you to enjoy the occasional bake as well.

▲ **Top left** The key to successful weight loss is to burn off more calories than you consume in a day, by eating the right foods and exercising regularly. **Top right** Coconut and coconut products are high in insoluble fibre and help you feel fuller for longer.

USING COCONUT SUGAR AS A SWEETENER

When trying to lose weight it is sensible to reduce the amount of sugar in your daily diet as far as possible. This is not just because of the calories sugar contains, but because sugar gives a short-lived energy boost, by raising blood sugar levels quickly. As speedily as blood sugar levels are raised, they then drop, causing energy levels to dip. This starts the whole cycle of wanting a quick energy-boost again. For maximum energy we need our blood sugar to remain fairly level throughout the day. While coconut sugar should not be eaten in large amounts, it is far preferable to white and brown refined sugars as it doesn't cause the same peaks and troughs in blood sugar levels. This is because it is a low-glycemic food (see page 19). In addition it provides useful vitamins and minerals in trace amounts including iron, zinc, potassium and calcium.

COCONUT FOR BEAUTY

It may surprise you that coconut is not only a superb addition to your diet, but is great as a beauty product too: it possesses outstanding benefits for your skin, hair, nails and teeth. Coconuts are already used in the beauty industry in moisturizers, body butters, bath oils, hair shampoos and conditioners. Even the ground shells are added to face and body exfoliants. Coconut is often blended with other less beneficial ingredients to make the product a certain consistency or emulsified with water to make the product go further.

In the world of beauty forget the adage that 'you get what you pay for'; there is no need to spend a fortune on products that come with fancy names and attractive packaging. Virgin coconut oil is not only free of artificial chemicals, colourings and preservatives, it also contains powerful antioxidants, vitamins C and E, and antibacterial properties. Refined coconut oil is a good alternative if you want a less expensive and scent-free product, but check the label, making sure it hasn't undergone chemical treatment that may have destroyed its healthy properties and that it doesn't contain undesirable additives.

▼ **Bottom left** Coconut oil is an excellent moisturizer for the skin. **Bottom right** You can use coconut oil on your nails as well as to treat dry skin on your feet.

COCONUT FOR HEALTHY SKIN

Our skin is the largest organ in the body. Eating well will help give it a healthy glow, but you also need to protect it on the outside. Coconut oil is an excellent moisturizer for your face, hands and body, especially if you have dry skin. It can't make you look any younger, but a fine coating will help restore lost moisture, and the antioxidants it contains will fight free-radical damage. This will help slow aging and improve skin elasticity.

Use coconut oil as a day or night cream; you won't need much as the oil will melt with the warmth of your fingertips so that you can apply it thinly. If you wish to make your own anti-aging product, gently heat 120ml/4fl oz/ ½ cup coconut oil until just melted and stir in 10ml/2 tsp pure vitamin E essential oil. If you occasionally suffer from spots, stir in 5 drops of pure tea tree essential oil instead of, or as well as, the vitamin E essential oil. Pour into a sterile or very clean and dry pot and leave to re-solidify. To keep skin plump, you should also make sure that you keep well hydrated.

Coconut oil can be used on the rest of your body too – it's especially good for rough elbows and knees and dry, cracked heels. It is also a natural and safe way to help prevent stretch marks during pregnancy and for treating baby's nappy rash and cradle cap.

COCONUT FOR GLOSSY HAIR

Exposure to sun and sea, regular swimming in a chlorinated pool, hair colouring, straightening and blow-drying can all damage your hair and may make it dry or brittle. Coconut oil is an effective and inexpensive conditioner that can revive your hair and help prevent further damage. The natural oils are quickly absorbed, smoothing the hair follicles and combating frizz.

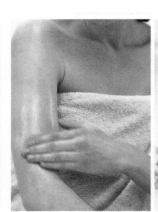

COCONUT FOR BEAUTY

▲ Comb coconut oil through damp hair for a cheap, effective and natural deep-conditioning treatment.

Simply apply a small amount to just the ends or driest parts, then shampoo as normal. For deeper conditioning, rinse your hair, then towel dry. Apply coconut oil to damp hair and use a wide-toothed comb to comb it through. Cover with a shower cap and wrap in a towel to keep warm, so that the oil penetrates the hair. Leave for up to an hour, then shampoo as normal.

COCONUT FOR YOUR NAILS

Rubbing a little coconut oil on your cuticles and nails daily will help moisturize them, making them more flexible and less likely to break. It will leave them with a glossy shine too. Coconut oil can also help prevent and even clear a fungal nail infection.

COCONUT FOR YOUR TEETH

Studies have shown there is a proven link between oral health and a healthy body. Gum (periodontal) disease can lead to health complications such as heart disease and strokes as the heart and other organs can become inflamed by bacterial endocarditis. Coconut oil is antibacterial, so is good for your teeth and mouth; extracts have already been added to some oral healthcare products.

OIL PULLING USING COCONUT OIL

The technique of swishing oil around your mouth is known as 'oil pulling' and is becoming an increasingly popular method of oral hygiene. The theory is that the oil draws out any bacteria that you can't reach with brushes or floss (hence the term 'pulling'). Using coconut oil can reduce the amount of streptococcus mutans, the main bacteria found in the mouth, which is a significant contributor to tooth decay.

Oil pulling is best done in the morning before you have had anything to eat or drink, although the evening is fine if you haven't eaten for an hour. You may feel queasy when you first try so you can use a smaller amount of oil until you get used to it. After a week or so, your teeth should feel noticeably cleaner and whiter.

1 Put 15ml/1 tbsp of virgin coconut oil in your mouth, then swish it around without swallowing it; it will soon become liquid.

2 Continue to slowly swish the oil around your mouth and through your teeth without swallowing. You will find that your saliva starts to dilute the oil a little. Keep going for as long as possible; take it slowly or your jaw may start to ache.

3 The first couple of times, it is fine to do it for just a few minutes, but you should build up to 15–20 minutes for the best results. If you have an urge to swallow or gag, spit out the oil and try again the next day.

4 At the end, the oil should be a milky colour and have a thinner consistency. Spit out the oil in a bin, rather than the sink or down the toilet. Rinse out your mouth with warm water, then brush your teeth.

OTHER USES FOR COCONUT

From clothes and shelter to agriculture and trade, coconuts are used not just as a source of food, but as a vital part of the economy and survival of many populations. Grown in over 90 countries, every part of the coconut palm can be used, from the tip of the highest leaf to the fibrous roots. The 'Coconut Palace' in Manila in the Philippines is a fantastic example of this versatility. Built almost entirely from coconut palms, it includes pillars made from the trunks, 101 coconut-shell chandeliers and a 24-seater dining table featuring more than 40,000 tiny pieces of inlaid coconut shell.

COIR

This is a natural fibre made from coconut husk, the individual fibre cells of which are hollow with thick walls made of cellulose. There are two types: brown coir and white coir. Brown coir is harvested from fully ripened coconuts and is thick and strong. It is used in doormats, sacking and brushes and as a stuffing fibre for mattresses. It is much stronger than cotton or flax, but is less flexible. White coir, which is pale brown, comes from green coconuts harvested after six months. The fibre is not as strong as brown coir, but is much more pliable and can be bleached and dyed to produce coloured yarn. It is generally spun to make rope or used to make yarn for rugs and mats.

Coir fibre is almost waterproof and is also resistant to sea water, making it ideal for coastal housing and for ship ropes, rigging and fishing nets. Coir is also used in horticulture as a potting compost, especially for orchids. In countries with heavy rain seasons and typhoons, coir is twined into rope and made into geotechnical netting that is then put on hilly slopes to prevent erosion. Grass seed can be put directly into the net and, after a few years, the nets biodegrade until all that is left is the grass that holds the hillside together.

The fibrous layer of the fruit is usually separated manually by driving the fruit down on to a spike to split it open. An experienced husker can split about 2,000 coconuts in a day. This can also be done by a machine, which crushes the fruit to open and remove the fibres.

COCONUT LEAVES

These provide material for roofing thatch, making bags, baskets and mats. The stiff midribs are often used for making brooms, while softer leaves are used to make 'ketupat', little rice dumplings that are enclosed in thin strips of woven leaves. Dried coconut leaves may be burnt to ash, which is then dug into soil to neutralize acidity and make it more workable.

COCONUT TRUNKS

Straight, strong and salt-resistant, coconut palm trunks are ideal for building houses, huts and bridges. Coconut timber is becoming increasingly popular in furniture-making as a substitute for non-sustainable hardwood. The hollowed-out trunk can also be used to make containers, canoes or drums.

◀ **Far left** Coconut shells can be used to make lovely lanterns. **Left** Using coconut's fibrous husk to make coir.

OTHER USES FOR COCONUT

▲ Coconut palms provide building material in the form of their strong trunks, used for supports, and their leaves, which can be woven to make walls and used for thatch.

COCONUT ROOTS

Growing just beneath the surface in a fibrous mass, coconut roots spread out to the same height as the tree. They can be boiled to produce a dye and are also used in mouthwashes as they have antibacterial properties, or used to make a basic toothbrush.

COCONUT SHELLS

These can be burnt as fuel and are a source of charcoal, effective for the removal of impurities in water filters, for example, or in refining sugar. The thin shells are wooden and make excellent bowls, which can be very basic, or beautifully carved ornamental ones. Buttons can also be made from the shells as well as simple musical instruments. The shells can be left rough or sanded smooth and polished, then have holes carefully drilled into them to make lanterns, or be carved into plant holders or little containers. Sometimes you'll find them in pet shops for use in a terrarium for reptiles or in an aquarium for fish. If you'd like to have a go at making something yourself with a shell, you could start with a simple bird feeder.

COCONUT BIRD FEEDER

Coconut shells make ideal vessels for containing bird seed, whether this is left loose or mixed with fat to make a high-energy snack that will help birds survive in cold weather.

½ coconut shell
strong weather-proof garden twine
½ packet suet (US chilled, grated shortening)
bird seed (the type used depends on which species of birds you find in your garden)
fresh unsalted peanuts, broken up
raisins

1 Carefully drill a hole in one end of the coconut shell, taking care to position it away from the cut edge of the coconut or it may split. Thread the garden twine through the hole.

2 Gently melt the suet in a pan, until it is clear and liquid. Remove the pan from the heat and add some bird seed, peanuts and raisins. Stir well to combine.

3 Pack the mixture into the coconut half. Use the other coconut half if you have some mixture left over, to make another bird feeder.

4 Leave to cool and set, then attach the feeder to a tree branch or a wooden post, angling the coconut upwards so that birds can easily access it.

RECIPES

Coconut water and coconut oil, along with many other coconut products, can easily be used in a wide range of delicious everyday recipes, from soups, snacks and salads to curries, casseroles and sweet treats. This section contains 70 simple dishes that you can try at home, along with cook's tips and variations that should inspire you to have a go at using these products in other favourite foods too. Just bear in mind that you should consider flavour combinations carefully and stick to the recommended daily intake quantities, then let your culinary creativity loose!

◀ Fragrant Thai Soup.
▼ Coconut Date Rolls.

DRINKS AND BREAKFASTS

COCONUT AND HAZELNUT SMOOTHIE

This nutty, creamy coconut drink should be sipped slowly. Sustaining and nourishing, it makes a good start to the day, or can be served as a nutritious snack-in-a-glass.

90g/3½oz/scant 1 cup whole blanched hazelnuts
10ml/2 tsp coconut sugar
2.5ml/½ tsp almond extract
150ml/¼ pint/⅔ cup coconut cream
200ml/7fl oz/1 cup coconut water
crushed ice

Makes 2 glasses

NUTRITIONAL INFORMATION:
Energy 581kcal/2404kJ; Protein 11.4g; Carbohydrate 12.2g, of which sugars 17.3g; Fat 54.6g, of which saturates 24.5g; Cholesterol 0mg; Calcium 74mg; Fibre 6.9g; Sodium 259mg.

1 Roughly chop the hazelnuts and lightly toast them in a small frying pan, turning frequently. Leave to cool, then tip the nuts into a blender or food processor with the sugar and blend well until finely ground.

2 Add the almond extract and coconut cream. Blend thoroughly until smooth. Strain the mixture through a sieve (strainer) into a jug or pitcher, pressing the pulp down with the back of a spoon to extract as much juice as possible. Stir in the coconut water.

3 Half-fill two glasses with crushed ice and pour over the coconut smoothie. Serve the drinks immediately, or chill in the refrigerator until ready to serve.

COCONUT AND PASSION FRUIT ICE

Few things beat the pure flavour of freshly juiced coconut. Blended with ice and teamed with passion fruit, this is a refreshing breakfast drink with a tropical taste.

1 fresh brown coconut
5ml/1 tsp sugar, preferably coconut sugar (optional)
150g/5oz crushed ice
3 passion fruit, pulp scooped out

Makes 2–3 glasses

NUTRITIONAL INFORMATION:
Energy 399kcal/1642kJ; Protein 4g; Carbohydrate 5g, of which sugars 5g; Fat 40.4g, of which saturates 35.2g; Cholesterol 0mg; Calcium 16mg; Fibre 11.6g; Sodium 22mg.

1 Drain the water from the coconut and break it open (see page 13). Remove the flesh, then pare off the brown skin. Push the flesh though a juicer with 150ml/¼ pint/⅔ cup water. Reserve the pulp for making coconut flour (see page 18), if liked. Stir in the sugar, if using.

2 Put the crushed ice in a blender or food processor and blend until slushy. Add the juiced coconut and drained coconut water. Process to just blend the ingredients. Pour into tall glasses, then, using a teaspoon, spoon the passion fruit on top.

COCONUT PORRIDGE WITH DATES AND NUTS

Full of valuable nutrients and fibre, puréed dates give a sweet flavour to this coconut porridge. Oats can help reduce blood cholesterol levels as part of a healthy diet.

250g/9oz/scant 2 cups fresh dates
225g/8oz/2 cups rolled oats
475ml/16fl oz/2 cups milk or coconut-flavoured or soya milk
300ml/½ pint/1¼ cups coconut water
a pinch of salt
50g/2oz/½ cup raw nuts, chopped

Serves 4

NUTRITIONAL INFORMATION:
Energy 450kcal/1897kJ; Protein 15.2g; Carbohydrate 66.8g, of which sugars 30.2g; Fat 15.4g, of which saturates 2.3g; Cholesterol 8mg; Calcium 188mg; Fibre 9.4g; Sodium 272mg.

1 Halve the dates and remove the stones (pits) and stems. Cover with boiling water and soak for 30 minutes, until softened. Strain, reserving 90ml/6 tbsp of the soaking water.

2 Remove the skin from the dates and purée them in a food processor with the soaking water.

3 Place the oats in a pan with the milk, coconut water and salt. Bring the mixture to the boil, then reduce the heat and simmer for 4–5 minutes until cooked, stirring frequently.

4 Serve topped with date purée and chopped nuts.

MIXED BERRY AND COCONUT QUINOA PORRIDGE

Made with a mixture of quinoa flakes and rolled oats, this coconut porridge is easy to prepare. The addition of fruit adds flavour and nutrients, especially vitamin C.

300ml/½ pint/1¼ cups milk or coconut-flavoured or soya milk
300ml/½ pint/1¼ cups coconut water
115g/4oz/1 cup quinoa flakes
50g/2oz/½ cup rolled oats
115g/4oz/1 cup mixed berries
coconut sugar, to serve

Serves 4–6

NUTRITIONAL INFORMATION:
Energy 125kcal/527kJ; Protein 6.5g; Carbohydrate 20.2g, of which sugars 7.6g; Fat 2.5g, of which saturates 0.7g; Cholesterol 4mg; Calcium 83mg; Fibre 3.1g; Sodium 166mg.

1 Put the milk, coconut water and quinoa flakes in a pan, bring to the boil and simmer for 5 minutes, until the flakes are softened.

2 Add the oats to the pan and simmer for a further 2 minutes, stirring often. Sprinkle the mixed berries over the top.

3 Cook the porridge for a further 2–3 minutes, until the berries are warmed through and just beginning to release their juices. Serve the porridge immediately, sprinkled with some coconut sugar and extra milk, if liked.

LUXURY COCONUT MUESLI

Commercially-made muesli (granola) can't compete with this home-made version, which is packed with seeds, grains, nuts and dried fruits, including coconut. This combination works well, or you can alter the balance of ingredients, if you like.

50g/2oz/½ cup sunflower seeds
25g/1oz/¼ cup pumpkin seeds
115g/4oz/1 cup hazelnuts
115g/4oz/1 cup rolled oats
115g/4oz/heaped 1 cup wheat flakes
115g/4oz/heaped 1 cup barley flakes
115g/4oz/1 cup raisins
115g/4oz/½ cup unsulphured dried apricots, chopped
50g/2oz/2 cups dried apple slices
50g/2oz/⅔ cup desiccated (dry unsweetened shredded) coconut
coconut water or coconut-flavoured milk, to serve

Serves 4

1 Put the sunflower and pumpkin seeds and the hazelnuts in a dry frying pan and cook over a medium heat for 3 minutes until golden, tossing regularly to prevent them burning. Chop the nuts.

2 Mix the toasted seeds and nuts with the remaining ingredients and leave to cool. Store in an airtight container. Serve the muesli with coconut water or coconut-flavoured milk.

Cook's tip
Dried fruit such as raisins, apricots and apples are naturally sweet, but if you like, sprinkle the muesli (granola) with a little coconut sugar or drizzle with coconut syrup before eating.

NUTRITIONAL INFORMATION: Energy 873kcal/3664kJ; Protein 21.8g; Carbohydrate 106.6g, of which sugars 40.1g; Fat 42.8g, of which saturates 9.2g; Cholesterol 0mg; Calcium 162mg; Fibre 19.1g; Sodium 75mg.

COCONUT GRANOLA

Nuts, seeds, oats and dried fruits baked with coconut syrup and coconut oil make a nutritious breakfast. Serve with coconut water or yogurt and some fresh fruit.

115g/4oz/1 cup rolled oats
115g/4oz/1 cup jumbo oats
50g/2oz/½ cup sunflower seeds
25g/1oz/2 tbsp sesame seeds
50g/2oz/½ cup hazelnuts, roasted
25g/1oz/¼ cup almonds, chopped
45ml/3 tbsp coconut oil
50ml/2fl oz/¼ cup coconut syrup or clear honey
50g/2oz/½ cup raisins
50g/2oz/½ cup dried cranberries
coconut water, or yogurt and fruit, to serve

Serves 4

1 Preheat the oven to 140°C/275°F/Gas 1. Mix together the oats, seeds and nuts in a large bowl.

2 Heat the coconut oil and syrup or honey in a large pan until melted, then remove the pan from the heat. Add the oat mixture to the pan and stir well to combine. Spread out on one or two baking sheets.

3 Bake for about 50 minutes until crisp, stirring occasionally. Remove from the oven and mix in the raisins and cranberries. Leave to cool, then store in an airtight container.

NUTRITIONAL INFORMATION: Energy 627kcal/2629kJ; Protein 15.1g; Carbohydrate 75.7g, of which sugars 27.5g; Fat 31.2g, of which saturates 9.6g; Cholesterol 0mg; Calcium 85mg; Fibre 8.9g; Sodium 37mg.

CRUNCHY COCONUT BREAKFAST MUFFINS

Toasted oat cereals are widely available – or you can use granola (see page 45) – and make a crunchy addition to these muffins. The raisins add a natural sweetness, so you don't need to add much sugar. Serve for brunch with a glass of coconut water.

150g/5oz/1¼ cups plain (all-purpose) flour
7.5ml/1½ tsp baking powder
30ml/2 tbsp coconut sugar
120ml/4fl oz/½ cup milk or coconut-flavoured or soya milk
120ml/4fl oz/½ cup coconut water
50g/2oz/¼ cup coconut oil, melted
1 egg
200g/7oz toasted oat cereal and raisins, mixed

Makes 10

Cook's tip
Make sure that the milk, coconut water and egg are all at room temperature.

1 Preheat the oven to 350°F/180°C/Gas 4. Lightly grease the cups of a muffin tin (pan) or line them with paper cases.

2 Sift the flour into a bowl. Add the baking powder, then the sugar and stir in. Make a well in the centre.

3 In a jug or pitcher, using a fork beat the milk and coconut water with the melted coconut oil and egg. Pour the liquid into the well in the flour mixture. Stir lightly until just combined.

4 Stir in the cereal and raisins. Bake for 20–25 minutes until risen and golden. Leave to cool in the tin for a few minutes, then turn out on to a wire rack to go completely cold. Serve fresh, or store in an airtight container for up to 3 days.

NUTRITIONAL INFORMATION: Energy 197kcal/828kJ; Protein 5g; Carbohydrate 26.1g, of which sugars 8.6g; Fat 8.8g, of which saturates 4.6g; Cholesterol 24mg; Calcium 51mg; Fibre 2.2g; Sodium 50mg.

COCONUT, LEMON AND RAISIN PANCAKES

Making pancakes is a great way to start the weekend. These ones are made with coconut water and coconut sugar and flavoured with lemon. Fried in coconut oil and topped with coconut syrup, they are a delicious way to start the day.

115g/4oz/1 cup self-raising (self-rising) flour
30ml/2 tbsp coconut sugar
1 egg
150ml/¼ pint/⅔ cup coconut water
grated rind of 1 lemon
coconut oil, for frying
25g/1oz/¼ cup raisins
coconut syrup or maple syrup, to serve

Makes 16

Variation
Substitute 15ml/1 tbsp coconut flour for 30ml/ 2 tbsp of the self-raising (self-rising) flour for more substantial pancakes.

1 Sift the flour into a bowl and stir in the sugar. Make a hollow in the middle. Crack the egg into a jug or pitcher and beat lightly. Stir in the coconut water and lemon rind. Add to the hollow and whisk to make a smooth, thick batter.

2 Put a little of the oil into a frying pan. Place the pan over a medium heat and, when hot, spoon in enough batter to form two or three pancakes, about 6–7cm/2½–3in in diameter.

3 Sprinkle a few raisins on each pancake. Keep the pan at a steady heat and when bubbles rise to the surface and burst, after 2–3 minutes, turn over with a palette knife. Cook for a further 2–3 minutes, until golden brown on the other side.

4 Remove the pancakes from the pan, place on a warm plate and keep warm by covering with a clean dish towel while you cook the remaining batter. Serve with coconut or maple syrup.

NUTRITIONAL INFORMATION: Energy 48kcal/203kJ; Protein 1.3g; Carbohydrate 8.5g, of which sugars 3.7g; Fat 1.2g, of which saturates 0.7g; Cholesterol 14mg; Calcium 29mg; Fibre 0.6g; Sodium 56mg.

MIXED VEGETABLE COCONUT OMELETTE

An omelette is a quick and easy impromptu brunch; here it is made mainly with leftovers such as potato and peas, all cooked together in coconut oil.

30ml/2 tbsp coconut oil
1 onion, finely chopped
1 garlic clove, crushed
1 or 2 fresh green chillies, chopped
fresh coriander (cilantro) sprigs, chopped, plus extra, to garnish
2.5ml/½ tsp ground cumin
1 firm tomato, chopped
1 small potato, cubed
25g/1oz/¼ cup cooked peas
25g/1oz/¼ cup cooked corn, or drained canned corn
2 eggs
25g/1oz/¼ cup grated Cheddar cheese
salt and ground black pepper, to taste

Serves 4–6

1 Heat the coconut oil in a large pan, then add the onion, garlic, chillies, fresh coriander, cumin, tomato, potato, peas and corn and fry for 2–3 minutes until they are well blended but the potato and tomato are still firm. Season to taste.

2 Increase the heat, beat the eggs and pour into the pan. Reduce the heat, cover the pan and cook until the bottom of the omelette is golden brown.

3 Sprinkle the omelette with the grated cheese. Place under a preheated hot grill (broiler) and cook until the egg sets and the cheese has melted.

4 Garnish the omelette with sprigs of fresh coriander and serve for a sustaining breakfast or brunch.

NUTRITIONAL INFORMATION: Energy 110kcal/456kJ; Protein 4.7g; Carbohydrate 5.9g, of which sugars 2.3g; Fat 7.7g, of which saturates 4.8g; Cholesterol 85mg; Calcium 52mg; Fibre 1.2g; Sodium 62mg.

COCONUT SCRAMBLED EGGS

If you want a breakfast that really wakes you up in the morning, these scrambled eggs are ideal, although they make a delicious light supper dish too.

45ml/3 tbsp coconut oil
1 large onion, finely chopped
1–2 fresh green chillies, finely chopped (seeded if preferred)
5ml/1 tsp grated fresh root ginger (optional)
2.5ml/½ tsp ground turmeric
2 salad tomatoes, finely chopped
15ml/1 tbsp chopped fresh coriander (cilantro) leaves
2.5ml/½ tsp salt, or to taste
4 large (US extra large) eggs
Indian bread or toast spread with coconut butter or oil, to serve

Serves 4

1 Heat the oil in a frying pan over a medium heat and add the onion, green chillies and ginger, if using. Stir-fry for 5–6 minutes, until the onion is soft.

2 Add the turmeric and chopped tomatoes to the pan and cook for 1 minute, then add the fresh coriander and salt.

3 Whisk the eggs then add them to the pan and whisk over the heat until they reach the desired consistency.

4 Transfer to plates and serve with warm Indian bread, or toast spread with coconut butter or coconut oil.

NUTRITIONAL INFORMATION: Energy 220kcal/911kJ; Protein 10.6g; Carbohydrate 6.6g, of which sugars 5.1g; Fat 17g, of which saturates 9.6g; Cholesterol 289mg; Calcium 70mg; Fibre 2.1g; Sodium 113mg.

SOUPS

CHILLED AVOCADO SOUP WITH COCONUT

This cold soup combines avocados with the distinctive flavours of onions, garlic, lemon and cumin and includes a healthy dose of coconut water. Avocado contains monounsaturated fat that can help to lower blood cholesterol levels.

3 ripe avocados
a bunch of spring onions (scallions), white parts only, trimmed and roughly chopped
2 garlic cloves, chopped
juice of 1 lemon
1.25ml/¼ tsp ground cumin
1.25ml/¼ tsp paprika
450ml/¾ pint/scant 2 cups fresh vegetable stock
300ml/½ pint/1¼ cups chilled coconut water
ground black pepper
chopped flat leaf parsley, to garnish

Serves 4

1 Starting well ahead to allow enough time for chilling, put the flesh of one avocado in a food processor or blender. Add the spring onions, garlic and lemon juice and purée until smooth. Add the second avocado and purée, then add the third, along with the spices and seasoning. Purée until smooth.

2 Gradually add the vegetable stock. Pour the soup into a bowl, cover with clear film (plastic wrap) and chill for 2–3 hours.

3 To serve, stir in the chilled coconut water, then season to taste with plenty of black pepper. Garnish with chopped parsley and serve immediately.

NUTRITIONAL INFORMATION: Energy 161kcal/665kJ; Protein 3.3g; Carbohydrate 3.7g, of which sugars 7g; Fat 14.9g, of which saturates 3.1g; Cholesterol 0mg; Calcium 44mg; Fibre 6.2g; Sodium 407mg.

SUMMER VEGETABLE AND COCONUT SOUP

This brightly coloured fresh-tasting tomato-based soup makes the most of summer vegetables in season. The subtle taste of coconut water adds just a hint of sweetness and brings all the flavours together to make this a memorable meal.

450g/1lb ripe plum tomatoes
225g/8oz ripe yellow tomatoes
30ml/2 tbsp coconut oil
1 large onion, finely chopped
15ml/1 tbsp tomato purée (paste)
225g/8oz courgettes (zucchini), trimmed and chopped
225g/8oz yellow courgettes, trimmed and chopped
3 waxy new potatoes, diced
2 garlic cloves, crushed
600ml/1 pint/2½ cups stock
475ml/16fl oz/2 cups coconut water
60ml/4 tbsp shredded fresh basil
50g/2oz/⅔ cup freshly grated Parmesan cheese
ground black pepper

Serves 4

1 Plunge all the tomatoes in a large heatproof bowl of boiling water for 30 seconds, refresh in ice-cold water, then peel away the skins and finely chop the flesh.

2 Heat the oil in a large pan and cook the onion for 5 minutes, until softened. Stir in the tomato purée, tomatoes, courgettes, potatoes and garlic. Mix well and cook gently for 10 minutes.

3 Pour in the stock and coconut water. Bring to the boil, lower the heat, half-cover the pan and simmer gently for 15 minutes, or until the vegetables are just tender. Add more stock or vegetable water if necessary.

4 Remove the pan from the heat and stir in the basil. Taste for seasoning and serve sprinked with Parmesan cheese.

NUTRITIONAL INFORMATION: Energy 226kcal/944kJ; Protein 12.3g; Carbohydrate 21.1g, of which sugars 18.4g; Fat 10.8g, of which saturates 7.5g; Cholesterol 28mg; Calcium 224mg; Fibre 9g; Sodium 629mg.

PUMPKIN SOUP WITH COCONUT

This simple puréed soup is a great cold-weather treat and makes a lovely appetizer, or a comforting lunch served with bread. It is especially good served with a swirl of coconut cream and a little coconut oil or melted butter drizzled over the top.

1kg/2¼lb pumpkin flesh, cubed
475ml/16fl oz/2 cups chicken or
  vegetable stock
475ml/16fl oz/2 cups coconut water
10ml/2 tsp coconut sugar
salt and ground black pepper,
  to taste
15g/½oz/1 tbsp coconut oil or butter
60ml/4 tbsp coconut cream

Serves 4

1 Put the pumpkin cubes into a pan with the stock and coconut water, and bring the liquid to the boil. Reduce the heat, cover the pan, and simmer for about 20 minutes, or until the pumpkin is tender.

2 Liquidize (blend) the soup in a blender, or use a potato masher to mash the flesh. Return the soup to the pan and bring it to the boil again.

3 Add the sugar to the pan and season to taste. Keep the pan over a low heat while you gently melt the coconut oil or butter in a small pan over a low heat.

4 Ladle the soup into serving bowls. Swirl a little coconut cream on to the surface and drizzle the oil or butter over the top. Serve immediately, offering extra coconut cream, so that you can enjoy the contrasting burst of sweet and creamy in each mouthful.

NUTRITIONAL INFORMATION: Energy 148kcal/616kJ; Protein 4.8g; Carbohydrate 9g, of which sugars 14.8g; Fat 10.5g, of which saturates 8.8g; Cholesterol 0mg; Calcium 76mg; Fibre 6.9g; Sodium 504mg.

FRAGRANT THAI FISH AND COCONUT SOUP

This light and aromatic soup is made with chunks of monkfish simmered in coconut water and a light stock, flavoured with lime, lemon grass, ginger, fresh herbs and chillies. An excellent source of protein, it is low in both fat and calories.

350ml/12fl oz/1½ cups coconut water
600ml/1 pint/2½ cups fish stock
4 lemon grass stalks
3 limes
2 small fresh hot red chillies, seeded and thinly sliced
2cm/¾in piece fresh root ginger, peeled and thinly sliced
6 fresh coriander (cilantro) stalks, with leaves
2 kaffir lime leaves, chopped
350g/12oz monkfish fillet, skinned and cut into 2.5cm/1in pieces
15ml/1 tbsp coconut vinegar
30ml/2 tbsp Thai fish sauce
30ml/2 tbsp chopped fresh coriander leaves, to garnish

Serves 4

1 Pour the coconut water and stock into a large pan and bring to the boil. Slice the bulb ends of the lemon grass diagonally into 3mm/⅛in-thick pieces. Peel off four wide strips of lime rind, avoiding the pith. Squeeze the limes and reserve the juice.

2 Add the lemon grass, lime rind, chillies, ginger, and fresh coriander stalks to the stock, with the kaffir lime leaves. Simmer gently for 1–2 minutes, then add the monkfish, coconut vinegar, Thai fish sauce and half the reserved lime juice.

3 Simmer for a further 3 minutes, until the fish is just tender; although cooked, it will still hold its shape.

4 Remove the coriander stalks from the pan and discard. Taste the broth and add more lime juice if necessary. Serve the soup very hot, sprinkled with the chopped fresh coriander leaves.

NUTRITIONAL INFORMATION: Energy 75kcal/317kJ; Protein 16.1g; Carbohydrate 1.3g, of which sugars 6.4g; Fat 0.6g, of which saturates 0.1g; Cholesterol 12mg; Calcium 11mg; Fibre 2.6g; Sodium 982mg.

COCONUT AND SEAFOOD SOUP

This delicious and attractive Thai soup is packed with coconut flavours, which all marry beautifully with the sweet seafood. Despite the long list of ingredients, it is extremely easy to put together and tastes wonderful.

300ml/½ pint/1¼ cups coconut water
300ml/½ pint/1¼ cups fish stock
5 thin slices fresh galangal or fresh root ginger
2 lemon grass stalks, chopped
3 kaffir lime leaves, shredded
a bunch garlic chives, about 25g/1oz
a small bunch fresh coriander (cilantro), about 15g/½oz
15ml/1 tbsp coconut oil
4 shallots, chopped
400ml/14fl oz can full-fat coconut milk
30–45ml/2–3 tbsp Thai fish sauce
45–60ml/3–4 tbsp Thai green curry paste
450g/1lb raw large prawns (shrimp), peeled and deveined
450g/1lb prepared squid
a little fresh lime juice
salt and ground black pepper
60ml/4 tbsp crisp-fried shallot slices, to serve

Serves 4

1 Pour the coconut water and stock into a large pan and add the galangal or ginger, lemon grass and half the kaffir lime leaves.

2 Reserve a few garlic chives for the garnish, then chop the remainder and add half to the pan. Strip the coriander leaves from the stalks and set the leaves aside. Add the stalks to the pan. Bring to the boil, reduce the heat to low and cover, then simmer gently for 20 minutes. Strain the stock into a bowl.

3 Rinse and dry the pan. Add the oil and shallots. Cook over a medium heat for 5–10 minutes, until the shallots are beginning to brown. Stir in the strained stock, coconut milk, the remaining kaffir lime leaves and 30ml/2 tbsp of the fish sauce. Heat gently until simmering and cook over a low heat for 5–10 minutes.

4 Stir in the curry paste and prawns, then cook for 3 minutes. Add the squid and cook for a further 2 minutes. Add the lime juice and add fish sauce to taste. Stir in the reserved chives and coriander leaves. Serve in bowls and sprinkle each with shallots.

NUTRITIONAL INFORMATION: Energy 265kcal/1117kJ; Protein 40.2g; Carbohydrate 8.4g, of which sugars 10.8g; Fat 8.2g, of which saturates 3.2g; Cholesterol 473mg; Calcium 170mg; Fibre 2.4g; Sodium 1347mg.

CHICKEN RICE SOUP WITH LEMON GRASS AND COCONUT

This wholesome soup, made with stock and coconut water, is light and refreshing, with the fragrant aroma of lemon grass. The rice and chicken are both easy to digest and this recipe makes a perfect pick-me-up if you have been under the weather.

1 small chicken or 2 chicken legs
600ml/1 pint/2½ cups coconut water
2 lemon grass stalks, trimmed, cut into 3 pieces, and lightly bruised with a rolling pin
15ml/1 tbsp Thai fish sauce
90g/3½oz/½ cup short grain rice, rinsed
1 bunch fresh coriander (cilantro) leaves, finely chopped, and 1 green or red chilli, seeded and cut into thin strips, to garnish
1 lime, cut in wedges, to serve
salt and ground black pepper

FOR THE STOCK
1 onion, quartered
2 cloves garlic, crushed
25g/1oz fresh root ginger, thinly sliced
2 lemon grass stalks, cut in half lengthways and bruised
2 dried red chillies
30ml/2 tbsp Thai dipping sauce

Serves 4

1 Put the chicken in a deep pan in which the chicken fits snugly. Add all the stock ingredients, then pour in 1.2 litre/2 pints/5 cups water. Bring to the boil for a few minutes, then reduce the heat and simmer gently with the lid on for 2 hours.

2 Skim off any fat from the stock, strain and reserve. Remove the skin from the chicken and shred the meat. Set aside.

3 Pour the stock back into the deep pan and add the coconut water. Bring to the boil. Reduce the heat and stir in the lemon grass stalks and Thai fish sauce. Stir in the rice and simmer, uncovered, for about 40 minutes.

4 Add the shredded chicken and season to taste. Ladle the piping hot soup into warmed individual bowls, garnish with chopped fresh coriander and the thin strips of chilli, and serve with lime wedges to squeeze over.

NUTRITIONAL INFORMATION: Energy 212kcal/890kJ; Protein 24.4g; Carbohydrate 18.3g, of which sugars 9.3g; Fat 4.5g, of which saturates 1.3g; Cholesterol 60mg; Calcium 30mg; Fibre 4.5g; Sodium 446mg.

TAMARIND PORK, VEGETABLE AND COCONUT SOUP

Known as 'Sinigang', this famous main-meal soup from the Philippines varies from region to region. It has a unique sharp and sour taste, which comes from tamarind. Coconut water adds a hint of sweetness that helps to balance the flavours.

600ml/1 pint/2½ cups coconut water
1.5 litres/2½ pints/6¼ cups pork or chicken stock
15–30ml/1–2 tbsp tamarind paste
30ml/2 tbsp patis (fish sauce)
25g/1oz fresh root ginger, grated
1 medium yam or sweet potato, cut into bitesize chunks
8–10 yard-long beans
225g/8oz kangkong (water spinach) or ordinary spinach, well rinsed
350g/12oz pork fillet (tenderloin), sliced widthways
2–3 spring onions (scallions), white parts only, finely sliced
ground black pepper

Serves 4–6

Cook's tip
Try to buy low-sodium stock (bouillon) cubes or fresh stock; the saltiness in this dish should come from the fish sauce and no additional salt is required.

1 In a wok or deep pan, bring the coconut water and the stock to the boil. Stir in the tamarind paste, patis and ginger, reduce the heat and simmer for about 20 minutes. Season the mixture with lots of ground black pepper.

2 Add the yam and beans to the wok or pan and cook gently for 3–4 minutes, until the yam is tender.

3 Stir in the spinach and the sliced pork and simmer gently for 2–3 minutes, until the pork is just cooked and turns opaque.

4 Ladle the soup into individual warmed bowls and sprinkle the sliced spring onions over the top. You will need chopsticks and a spoon to eat with.

NUTRITIONAL INFORMATION: Energy 154kcal/652kJ; Protein 17.6g; Carbohydrate 14.5g, of which sugars 13.5g; Fat 3.2g, of which saturates 0.9g; Cholesterol 37mg; Calcium 100mg; Fibre 6.5g; Sodium 986mg.

INDIAN LAMB SOUP WITH RICE AND COCONUT

Coconut milk adds a satisfying creamy richness to this meaty soup. Thickened with long grain rice and flavoured with cumin and coriander seeds, the recipe is based on the classic Indian mulligatawny soup. Toasted coconut adds the finishing flourish.

2 onions, chopped
6 garlic cloves, crushed
5cm/2in fresh root ginger, grated
45ml/3 tbsp coconut oil
30ml/2 tbsp black poppy seeds
5ml/1 tsp cumin seeds
5ml/1 tsp coriander seeds
2.5ml/½ tsp ground turmeric
450g/1lb boneless lamb chump chops, trimmed and cut into bitesize pieces
1.25ml/¼ tsp cayenne pepper
600ml/1 pint/2½ cups coconut water
600ml/1 pint/2½ cups lamb stock
50g/2oz/⅓ cup long grain rice
30ml/2 tbsp lemon juice
60ml/4 tbsp coconut milk
salt and ground black pepper
fresh coriander (cilantro) sprigs and toasted flaked coconut, to garnish

Serves 6

1 Process the onions, garlic, ginger and 15ml/1 tbsp of the oil in a food processor or blender to form a paste. Set aside.

2 Heat a small, heavy frying pan. Add the poppy, cumin and coriander seeds and toast for a few seconds, shaking the pan, until they begin to release their aroma. Transfer to a mortar and grind them to a powder. Stir in the turmeric. Set aside.

3 Heat the remaining oil in a heavy pan. Fry the lamb in batches over a high heat for 4–5 minutes until browned. Remove the lamb. Add the onion paste and cook for 2 minutes, stirring continuously. Stir in the spices and cook for 1 minute. Return the meat to the pan with any meat juices. Add the cayenne, coconut water, stock and seasoning. Bring to the boil, cover and simmer for 30 minutes or until the lamb is tender.

4 Stir in the rice, then cover and cook for a further 15 minutes. Add the lemon juice and coconut milk and simmer for a further 2 minutes. Serve garnished with fresh coriander and coconut.

NUTRITIONAL INFORMATION: Energy 222kcal/926kJ; Protein 19g; Carbohydrate 11.1g, of which sugars 9.3g; Fat 12.2g, of which saturates 7.4g; Cholesterol 56mg; Calcium 47mg; Fibre 3.9g; Sodium 490mg.

SNACKS AND SALADS

COCONUT GUACAMOLE

This is a coconut twist on a dish that is a classic of Tex-Mex cooking. Usually served as an appetizer with vegetable sticks or corn chips for dipping, it makes a healthy snack, simple lunch or accompaniment for fish or meat.

2 large ripe avocados
1 small red onion, finely chopped
1 fresh chilli, seeded and chopped
½–1 garlic clove, crushed with a
    little salt (optional)
finely shredded rind of ½ lime and
    juice of 1–1½ limes
a pinch of coconut sugar
225g/8oz tomatoes
30ml/2 tbsp chopped coriander
    (cilantro)
2.5–5ml/½–1 tsp ground cumin
15ml/1 tbsp coconut oil, melted
15–30ml/1–2 tbsp coconut cream
salt and ground black pepper
lime wedges, sea salt, and fresh
    coriander, to garnish

Serves 4

1 Halve, stone (pit) and peel the avocados. Set one half aside and roughly mash the remainder in a bowl using a fork.

2 Add the onion, chilli, garlic (if using), lime rind, juice of 1 lime and the sugar. Seed and finely chop the tomatoes, then stir in with the coriander. Add ground cumin, seasoning and more lime juice to taste. Stir in the coconut oil.

3 Dice the remaining avocado and stir the pieces into the guacamole, then cover with clear film (plastic wrap) and leave to stand for at least 15 minutes so the flavour develops.

4 Stir in the coconut cream. Serve immediately with lime wedges dipped in sea salt, and fresh coriander sprigs.

NUTRITIONAL INFORMATION: Energy 295kcal/1219kJ; Protein 3.4g; Carbohydrate 7.4g, of which sugars 4.8g; Fat 28.2g, of which saturates 8.6g; Cholesterol 0mg; Calcium 35mg; Fibre 7.3g; Sodium 15mg.

COCONUT AND HERB CHILLI DIP

Serve this mild Colombian chilli dip with empanadas or crudités. The dip is made in large quantities and kept in the refrigerator for use as and when it is needed.

1 small habañero chilli, seeded
1–2 spring onions (scallions)
1 tomato, seeded and skinned
90ml/6 tbsp chopped fresh
    coriander (cilantro)
75ml/5 tbsp coconut vinegar
75ml/5 tbsp coconut water
10ml/2 tsp lime juice
5ml/1 tsp salt
7.5ml/½ tbsp coconut oil

Makes 150ml/¼ pint/1¼ cups

1 Very finely chop the chilli, spring onions, tomato and fresh coriander. Combine in a large bowl, then add all the remaining ingredients and stir well to combine thoroughly.

2 Decant as much as you want for one serving into a serving bowl, cover with clear film (plastic wrap) and set aside at room temperature for at least 1 hour before serving, so that the flavours can blend. Store the remainder, covered, in the refrigerator.

NUTRITIONAL INFORMATION: Energy 72kcal/299kJ; Protein 4.5g; Carbohydrate 5.6g, of which sugars 9.9g; Fat 3.7g, of which saturates 2.7g; Cholesterol 0mg; Calcium 70mg; Fibre 3.5g; Sodium 2181mg.

GRIDDLED COCONUT POLENTA WITH TANGY PEBRE

Polenta is an excellent wheat-free starchy carbohydrate, here flavoured with chillies, fresh herbs and coconut and served with pebre, a spicy Chilean condiment.

10ml/2 tsp crushed dried chilli flakes
600ml/1 pint/2½ cups coconut water
750ml/1¼ pints/3 cups mild vegetable stock or water
250g/9oz/1¼ cups quick-cook polenta
45ml/3 tbsp coconut oil
30ml/2 tbsp chopped fresh dill
30ml/2 tbsp chopped fresh coriander (cilantro)

FOR THE PEBRE
½ red onion, finely chopped
4 drained bottled sweet cherry peppers, finely chopped
1 fresh medium-hot red chilli, seeded and finely chopped
1 small red (bell) pepper, quartered, seeded and diced
10ml/2 tsp coconut vinegar or cider vinegar
30ml/2 tbsp coconut oil
4 tomatoes, cored and chopped
45ml/3 tbsp chopped fresh coriander (cilantro)

Serves 6

1 Put the chilli flakes in a pan with the coconut water and stock and bring to the boil. Pour in the polenta in a steady stream, whisking all the time. Reduce the heat and continue to whisk for a few minutes. When the polenta is thick, whisk in half the oil and the herbs. Pour into a greased 33 x 23cm/13 x 9in baking tray and leave to cool. Chill, uncovered, overnight.

2 To make the pebre, place the onion, cherry peppers, chilli and diced pepper in a mortar with the coconut vinegar and coconut oil. Pound with a pestle for 1 minute, then tip into a dish. Stir in the tomatoes and coriander. Cover and leave in a cool place.

3 Bring the polenta to room temperature. Cut into 12 even triangles and brush the top with the remaining coconut oil.

4 Heat a griddle until drops of water sprinkled on the surface evaporate instantly. Lower the heat to medium and grill (broil) the triangles in batches oiled-side down for 2 minutes, then turn through 180 degrees and cook for 1 minute more, to achieve a checked effect. Serve immediately, with the pebre.

NUTRITIONAL INFORMATION: Energy 273kcal/1139kJ; Protein 7.3g; Carbohydrate 35.8g, of which sugars 11g; Fat 11g, of which saturates 8g; Cholesterol 0mg; Calcium 28mg; Fibre 6.1g; Sodium 264mg.

COCONUT PATTIES

In India, these little morsels are made from grated fresh coconut. Use this if you have time, but for simplicity, here they are made with shredded coconut softened with coconut water to replenish the moisture.

150g/5oz/2 cups desiccated (dry unsweetened shredded) coconut
150ml/¼ pint/⅔ cup coconut water, heated until warm
2 slices of bread, crusts removed
75g/3oz/⅔ cup gram flour (besan)
1–3 fresh green chillies, roughly chopped (seeded if preferred)
2.5cm/1in piece of fresh root ginger, peeled and roughly chopped
1 large garlic clove, peeled and roughly chopped
15ml/1 tbsp fresh coriander (cilantro) leaves and stalks, chopped
2.5ml/½ tsp chilli powder, or to taste
3.75ml/¾ tsp salt, or to taste
1 medium onion, finely chopped
coconut oil, for shallow-frying
chutney, to serve

Makes 16

1 Put the coconut in a large mixing bowl and pour over the warm coconut water. Set aside for 10 minutes for the coconut to absorb the water. Cut the bread into small pieces.

2 Place all the ingredients, except the onion and the oil, in a food processor and blitz to a smooth paste. Transfer the mixture to a bowl and add the onion. Mix thoroughly and divide the mixture into 16 balls, each the size of a lime. Flatten the coconut balls to form 16 smooth, round patties. If the mixture sticks to your fingers, moisten your palms with water between patties.

3 Put 15ml/1 tbsp coconut oil in a non-stick frying pan and heat over a medium-high heat. Fry the patties in batches for 3–4 minutes on each side, until browned all over.

4 Lift out and drain on kitchen paper. Keep warm while you fry the other patties, adding more oil to the pan when necessary.

NUTRITIONAL INFORMATION: Energy 109kcal/451kJ; Protein 2.2g; Carbohydrate 5.3g, of which sugars 2.2g; Fat 8.9g, of which saturates 7.4g; Cholesterol 0mg; Calcium 22mg; Fibre 3g; Sodium 44mg.

TOFU FALAFELS WITH COCONUT

Traditionally made from chickpeas, this version of the popular snack falafels uses tofu and coconut oil as its base. These crunchy balls make a great lunch served with pitta bread and a dip, such as hummus or sweet chilli sauce.

45ml/3 tbsp coconut oil
2 large onions, finely chopped
3 garlic cloves, crushed
500g/1¼lb firm tofu, drained and processed to a smooth paste
200g/7oz/3¾ cups fresh breadcrumbs
15g/½oz bunch fresh parsley, finely chopped
45ml/3 tbsp coconut aminos or soy sauce
50g/2oz/4 tbsp sesame seeds, toasted
5ml/1 tsp ground cumin
15ml/1 tbsp ground turmeric
60ml/4 tbsp tahini
juice of 1 lemon
1.25ml/¼ tsp cayenne pepper
warmed pitta bread and some hummus or a sweet chilli sauce, to serve

Serves 4–6

Cook's tip
If the falafals are quite sticky when shaping, roll them in a little coconut flour.

1 Heat 30ml/2 tbsp of the coconut oil in a large frying pan and sauté the onion and garlic over a medium heat for 2–3 minutes, until softened. Set aside to cool slightly.

2 Preheat the oven to 180°C/350°F/Gas 4. In a bowl, combine the remaining ingredients and oil, then stir in the onion mixture.

3 Form the mixture into 2.5cm/1in-diameter balls and place them on an oiled baking sheet. Bake for 30 minutes, or until the balls are crusty on the outside but still moist on the inside.

4 Spear each of the hot falafels with a cocktail stick (toothpick), and serve immediately with warmed pitta bread and some hummus or sweet chilli sauce.

NUTRITIONAL INFORMATION: Energy 323kcal/1345kJ; Protein 14.6g; Carbohydrate 21.8g, of which sugars 6.5g; Fat 21.2g, of which saturates 7.1g; Cholesterol 0mg; Calcium 670mg; Fibre 5.3g; Sodium 697mg.

THAI COCONUT RICE SALAD

Lemon grass, chilli, coconut and almonds add exciting tastes and textures to this colourful main-course salad. It is perfect for a relaxed lunch as it can be mostly prepared ahead. Just before serving, add the avocado, basil leaves and dressing.

350g/12oz/3 cups cooked rice
1 Asian pear, cored and diced
50g/2oz dried shrimp, chopped
1 avocado, peeled, stoned (pitted) and diced
½ medium cucumber, finely diced
2 lemon grass stalks, finely chopped
30ml/2 tbsp sweet chilli sauce
1 fresh chilli, seeded and finely sliced
115g/4oz/1 cup flaked (sliced) almonds, toasted
a small bunch fresh coriander (cilantro), chopped
Thai sweet basil leaves, to garnish

FOR THE DRESSING
300ml/½ pint/1¼ cups coconut water
10ml/2 tsp shrimp paste
15ml/1 tbsp coconut sugar
2 fresh kaffir lime leaves, torn into small pieces
½ lemon grass stalk, sliced

Serves 4–6

Variation
Use whatever fresh fruit, vegetables and even leftover meat that you might have to make this salad.

1 Make the dressing. Put the coconut water in a small pan with the shrimp paste, coconut sugar, kaffir lime leaves and lemon grass. Heat gently, stirring, until the sugar dissolves, then bring to boiling point and simmer for 5 minutes. Strain into a bowl and set aside until cold.

2 Put the cooked rice in a large salad bowl and fluff up the grains with a fork. Add the Asian pear, dried shrimp, avocado, cucumber, lemon grass and sweet chilli sauce. Mix well.

3 Add the diced chilli, almonds and coriander to the bowl and toss well. Garnish with Thai basil leaves and serve with the bowl of dressing to spoon over the top of individual portions.

NUTRITIONAL INFORMATION: Energy 302kcal/1261kJ; Protein 13.2g; Carbohydrate 26.5g, of which sugars 10.6g; Fat 16.6g, of which saturates 2.1g; Cholesterol 51mg; Calcium 198mg; Fibre 3.8g; Sodium 699mg.

GADO GADO SALAD WITH COCONUT

This well-known Indonesian salad combines steamed and fresh vegetables with hard-boiled eggs. Coconut water adds extra nuttiness and a sweet touch to the richly flavoured dressing made from peanuts and coconut aminos.

225g/8oz new potatoes, halved
2 carrots, cut into sticks
115g/4oz green beans
½ cauliflower, broken into florets
¼ firm white cabbage, shredded
200g/7oz bean or lentil sprouts
4 eggs, hard-boiled and quartered
a bunch of watercress (optional)

FOR THE SAUCE
90ml/6 tbsp crunchy peanut butter
300ml/½ pint/1¼ cups coconut water
1 garlic clove, crushed
15ml/1 tbsp dry sherry
15ml/1 tbsp coconut vinegar
5ml/1 tsp anchovy essence (extract)
30ml/2 tbsp coconut aminos or dark soy sauce
10ml/2 tsp coconut sugar

Serves 6

Variation
For a richer sauce, you can use coconut milk instead of coconut water or add 15ml/ 1 tbsp coconut cream.

1 Place the halved potatoes in a metal colander or steamer and set over a pan of gently boiling water. Cover the pan or steamer with a lid and cook the potatoes for 10 minutes.

2 Add the rest of the vegetables to the steamer and steam for a further 10 minutes, until tender.

3 Cool and arrange on a platter with the egg quarters and the watercress, if using.

4 Beat together the peanut butter, coconut water, garlic, sherry, coconut vinegar, anchovy essence, coconut aminos and sugar in a large bowl until smooth. Drizzle a little sauce over each portion then pour the rest into a small bowl and serve separately.

NUTRITIONAL INFORMATION: Energy 273kcal/1145kJ; Protein 10g; Carbohydrate 40g, of which sugars 15g; Fat 8g, of which saturates 4g; Cholesterol 19mg; Calcium 209mg; Fibre 4g; Sodium 70mg.

CITRUS AND COCONUT CHICKEN COLESLAW

This zesty coleslaw makes a refreshing change from the rich mayonnaise version. Crisp vegetables are combined with chicken fried in coconut oil and a citrus dressing to make a nutritious salad that is ample enough to be served as a main course.

90ml/6 tbsp coconut oil
6 boneless chicken breast fillets, skinned
4 oranges
2 limes
5ml/1 tsp Dijon mustard
15ml/1 tbsp clear honey
300g/11oz/2¾ cups white cabbage, finely shredded
300g/11oz carrots, peeled and finely sliced
2 spring onions (scallions), sliced
2 celery sticks, cut into matchsticks
30ml/2 tbsp fresh tarragon, chopped
salt and ground black pepper

Serves 6

Variation
For a creamier result, mayonnaise, crème fraîche or coconut cream can be used to dress the chicken instead of the orange, oil and honey mixture.

1 Heat 30ml/2 tbsp of the oil in a large, heavy frying pan. Add the chicken and cook for 15–20 minutes, or until cooked through and golden brown. Remove from the pan and leave to cool.

2 Peel two of the oranges and the limes, cut off the pith, then separate the segments and set aside. Grate the rind and squeeze the juice from one of the remaining oranges and place in a pan.

3 Add the remaining oil to the pan with the Dijon mustard and honey and season with salt and pepper. Gently heat until the oil has melted. Remove from the heat and whisk well.

4 Put the cabbage, carrots, spring onions and celery in a bowl and pour over two-thirds of the dressing. Leave to stand for 10 minutes.

5 Squeeze the juice from the rest of the oranges and mix into the remaining dressing with the orange and lime segments and tarragon. Slice the chicken and stir into the orange mixture. Spoon the salad on to plates and serve.

NUTRITIONAL INFORMATION: Energy 312kcal/1308kJ; Protein 32.6g; Carbohydrate 17.1g, of which sugars 16.5g; Fat 13g, of which saturates 10g; Cholesterol 88mg; Calcium 113mg; Fibre 5.6g; Sodium 135mg.

MAIN COURSES

BEEF, MUSHROOM AND COCONUT SALAD

This simple yet substantial salad combines tender thin strips of grilled beef with mushrooms cooked in heart-healthy coconut oil and fresh salad vegetables. A zesty lime dressing complements the robust flavours.

500g/1¼lb fillet (tenderloin) or rump (round) steak
30ml/2 tbsp coconut oil
2 small mild red chillies, seeded and sliced
225g/8oz/3¼ cups fresh shiitake mushrooms, stems removed and caps sliced
1 cos or romaine lettuce, torn into strips
175g/6oz cherry tomatoes, halved
5cm/2in piece cucumber, peeled, halved and thinly sliced
45ml/3 tbsp toasted sesame seeds

FOR THE DRESSING
3 spring onions (scallions), chopped
2 garlic cloves, finely chopped
juice of 1 lime
30ml/2 tbsp chopped fresh coriander (cilantro)
15–30ml/1–2 tbsp Thai fish sauce
5ml/1 tsp coconut sugar

Serves 4

Variation
Instead of beef, make this with 450g/1lb lamb neck fillet, trimmed of fat and cut in half lengthways. Grill (broil) for 6–7 minutes on each side. Cool and slice thinly.

1 Preheat the grill (broiler) to medium, then cook the steak for 2–4 minutes on each side, depending on how well done you like it. Leave to cool for at least 15 minutes. Slice the meat as thinly as possible and place the slices in a bowl.

2 Heat the coconut oil in a small frying pan. Add the seeded and sliced red chillies and the sliced shiitake mushroom caps. Cook for 5 minutes, stirring occasionally.

3 Turn off the heat and add the steak slices to the pan. Stir well to coat the beef slices in the chilli and mushroom mixture.

4 Make the dressing by mixing all the ingredients together in a bowl, then pour it over the meat mixture and toss gently.

5 Arrange the lettuce, tomatoes and cucumber on a serving plate. Spoon the steak mixture in the centre and sprinkle the sesame seeds over the top. Serve immediately.

NUTRITIONAL INFORMATION: Energy 244kcal/1018kJ; Protein 30.8g; Carbohydrate 3.8g, of which sugars 3.8g; Fat 11.7g, of which saturates 7.1g; Cholesterol 73mg; Calcium 57mg; Fibre 2g; Sodium 400mg.

VEGETABLE STIR-FRY WITH COCONUT

The contrast between the crunchy seeds and vegetables and the rich, savoury sauce made with coconut aminos, sugar and vinegar is what makes this dish so delicious. Seeds are nutritional powerhouses, packed with beneficial oils and protein.

30ml/2 tbsp coconut oil
30ml/2 tbsp sesame seeds
30ml/2 tbsp sunflower seeds
30ml/2 tbsp pumpkin seeds
2 garlic cloves, finely chopped
2.5cm/1in piece fresh root ginger, peeled and finely chopped
2 large carrots, cut into matchsticks
2 large courgettes (zucchini), cut into matchsticks
90g/3½oz/1½ cups oyster mushrooms, torn into pieces
150g/5oz watercress or baby spinach leaves, coarsely chopped
a small bunch fresh mint or coriander (cilantro), chopped
60ml/4 tbsp black bean sauce
30ml/2 tbsp coconut aminos or light soy sauce
15ml/1 tbsp coconut sugar or light muscovado (brown) sugar
30ml/2 tbsp coconut vinegar or rice vinegar

Serves 4

1 Heat the oil in a wok or large frying pan. Add the seeds. Toss over a medium heat for 1 minute, then add the garlic and ginger and continue to stir-fry for a further minute. Do not let the garlic burn or it will taste bitter.

2 Add the carrot and courgette matchsticks and the sliced mushrooms to the wok and stir-fry for a further 5 minutes, or until all the vegetables are crisp-tender and golden at the edges.

3 Add the watercress or spinach with the fresh herbs. Toss together over the heat for 1 minute, then stir in the black bean sauce, coconut aminos or soy sauce, sugar and vinegar. Stir-fry for 1–2 minutes, until combined and hot. Serve immediately.

NUTRITIONAL INFORMATION: Energy 257kcal/1065kJ; Protein 8.8g; Carbohydrate 11.3g, of which sugars 10.7g; Fat 19.9g, of which saturates 7.4g; Cholesterol 0mg; Calcium 275mg; Fibre 6.2g; Sodium 972mg.

SPICY BLACK BEAN COCONUT BURGERS

These vegetarian burgers use low-fat and fibre-rich canned black beans – a handy kitchen standby all year round. Super-healthy quinoa cooked in coconut water and nuts combine with jalapeño peppers, chilli, lime and herbs to make these tasty treats.

115g/4oz/⅔ cup pearl quinoa
350ml/12fl oz/1½ cups coconut water
30ml/2 tbsp coconut oil
1 medium onion, finely chopped
1 stick celery, finely chopped
2 cloves garlic, crushed
6 jalapeño peppers, finely chopped
1 red or green chilli, finely chopped
2 medium carrots, peeled and grated
75g/3oz/½ cup roasted peanuts
1 lime, rind and juice
15ml/1 tbsp fresh coriander (cilantro), roughly chopped
400g/14oz can black beans, drained and rinsed
15ml/1 tbsp coconut flour or plain (all-purpose) flour, for shaping
salt and ground black pepper
burger buns, shredded lettuce, sliced tomato and crème fraîche, to serve

Serves 6

1 Simmer the quinoa in the coconut water for 15–17 minutes until soft.

2 In another pan heat 15ml/1 tbsp of the oil and add the onion, celery, garlic, jalapeños, chilli and salt and pepper. Cook for 2–3 minutes on medium heat, then add the carrot and cook for 3 minutes. Leave to cool.

3 Blitz the quinoa, peanuts, lime juice and rind, and coriander in a food processor. Add the vegetable mixture, salt and pepper. Pulse until blended. Shape into six burgers, using flour as required.

4 Fry the burgers in the remaining 15ml/1 tbsp oil, adding a little more if needed, and turning them halfway through cooking. Serve in halved buns with lettuce, tomato and crème fraîche.

NUTRITIONAL INFORMATION: Energy 251kcal/1054kJ; Protein 12.3g; Carbohydrate 27.6g, of which sugars 8.8g; Fat 10.9g, of which saturates 4.5g; Cholesterol 0mg; Calcium 50mg; Fibre 4.8g; Sodium 154mg.

BARLEY COCONUT RISOTTO WITH SQUASH AND LEEKS

This is more like a pilaff made with pearl barley and coconut water than a classic risotto. Sweet leeks and roasted butternut squash are superb with this earthy grain.

200g/7oz/1 cup pearl barley
1 butternut squash, peeled, seeded and cut into chunks
10ml/2 tsp chopped fresh thyme
60ml/4 tbsp coconut oil
4 leeks, cut into fairly thick diagonal slices
2 garlic cloves, finely chopped
175g/6oz/2½ cups brown cap (cremini) mushrooms, sliced
2 carrots, coarsely grated
about 120ml/4fl oz/½ cup coconut water
30ml/2 tbsp chopped fresh flat leaf parsley
45ml/3 tbsp pumpkin seeds, toasted, or chopped walnuts
ground black pepper

Serves 4

1 Rinse the barley, then cook it in simmering water, keeping the pan part-covered, for 35–45 minutes, or until tender. Drain. Preheat the oven to 200°C/400°F/Gas 6.

2 Put the squash in a roasting pan with half the thyme and ground pepper. Melt 30ml/2 tbsp of the oil and drizzle over, mixing to coat. Roast for 30 minutes, stirring once, until tender.

3 Heat the remaining coconut oil in a large frying pan. Cook the leeks and garlic gently for 5 minutes. Add the mushrooms and remaining thyme, then cook until the liquid from the mushrooms evaporates and they begin to fry.

4 Stir in the carrots and cook for about 2 minutes, then add the barley and most of the coconut water. Season well and partially cover the pan. Cook for a further 5 minutes. Pour in the remaining coconut water if the mixture seems dry.

5 Stir in the squash and the parsley. Season to taste and serve sprinkled with the toasted pumpkin seeds or walnuts.

NUTRITIONAL INFORMATION: Energy 499kcal/2105kJ; Protein 14.4g; Carbohydrate 73.3g, of which sugars 20.1g; Fat 18.6g, of which saturates 10.7g; Cholesterol 0mg; Calcium 204mg; Fibre 14.7g; Sodium 107mg.

CHICKPEA COCONUT PILAU

In this colourful pilau, basmati rice cooked in coconut water is spiced with cinnamon, cardamom and cloves and combined with chickpeas, fresh coriander and mint.

225g/8oz/1 cup basmati rice
45ml/3 tbsp coconut oil
1 large onion, finely sliced
2.5cm/1in piece cinnamon stick
6 green cardamom pods, bruised
6 cloves
10–12 black peppercorns
5ml/1 tsp ground coriander
2.5ml/½ tsp ground cumin
2.5ml/½ tsp chilli powder
2.5ml/½ tsp ground turmeric
2 tomatoes, chopped
400g/14oz canned chickpeas, drained and rinsed
5ml/1 tsp salt, or to taste
350ml/12fl oz/1½ cups coconut water
30ml/2 tbsp finely chopped fresh coriander (cilantro)
15ml/1 tbsp finely shredded fresh mint
raita or vegetable curry, to serve

Serves 6

Cook's tip
Pulses (legumes) such as chickpeas have a low GI (Glycaemic Index) factor, so they release energy slowly into the bloodstream. This can help to control hunger and appetite.

1 Wash the rice in several changes of water. Transfer to a bowl, cover with more cold water and soak for 20 minutes. Drain.

2 Melt the coconut oil in a heavy frying pan over a low-medium heat. Add the onion and fry for 6 minutes, until well browned, stirring frequently. Remove from the pan with a slotted spoon, leaving as much oil as possible behind. Set aside.

3 Add the cinnamon, cardamom, cloves and black peppercorns to the pan and cook for 30 seconds, then stir in the coriander, cumin, chilli powder and turmeric.

4 Cook for 1 minute, then add the tomatoes, chickpeas, rice and salt. Add the coconut water and 200ml/7fl oz/scant 1 cup boiling water. Bring to the boil and let it cook, uncovered, for 2 minutes. Reduce the heat to low, cover, and cook for 10 minutes. Switch off the heat. Leave to stand undisturbed for 15 minutes.

5 Add the fresh coriander and mint, and fluff up the rice with a fork. Transfer to a serving dish, garnish with the fried onion and serve with raita or a vegetable curry.

NUTRITIONAL INFORMATION: Energy 279kcal/1166kJ; Protein 9g; Carbohydrate 43.6g, of which sugars 6.9g; Fat 7.8g, of which saturates 5g; Cholesterol 0mg; Calcium 63mg; Fibre 5.6g; Sodium 263mg.

QUINOA AND COCONUT CURRY WITH SEEDED FLATBREAD

A nutritious grain from South America, quinoa makes a great alternative to rice. Here it is combined with coconut water, lentils, cauliflower, carrots, spinach and tomatoes for a super-healthy meal. Lime pickle gives the curry a spicy kick. The seeded flatbreads are made from a mixture of whole-wheat and quinoa flour and cooked in a frying pan.

15ml/1 tbsp coconut oil
1 small onion, finely sliced
1 clove garlic, peeled and crushed
15ml/1 tbsp garam masala
115g/4oz cauliflower florets
1 medium carrot, peeled and diced
125g/4¼oz/¾ cup pearl quinoa
150g/5oz/generous ½ cup red lentils
475ml/16fl oz/2 cups coconut water
250ml/8fl oz/1 cup boiling water
400g/14oz canned chopped tomatoes
15ml/1 tbsp tomato purée (paste)
115g/4oz spinach, washed and shredded
15ml/1 tbsp fresh coriander (cilantro), chopped, plus extra for garnishing
30ml/2 tbsp lime pickle
salt and ground black pepper
lime wedges and lime pickle, to serve

FOR THE SEEDED FLATBREADS
30ml/2 tbsp mixed seeds, such as pumpkin, sunflower and poppy
175g/6oz/1½ cups wholemeal (whole-wheat) flour
175g/6oz/1½ cups quinoa flour
7.5ml/1½ tsp baking powder
10ml/2 tsp ground cumin
2.5ml/½ tsp salt
30ml/2 tbsp coconut oil
about 175ml/6fl oz/¾ cup coconut water, at room temperature
coconut oil, for frying

Serves 4

1 Make the flatbread dough. Toast the seeds for 5 minutes under a medium grill (broiler) or in a dry frying pan.

2 Combine the seeds with the flours, baking powder, cumin and salt in a large bowl. Add the oil and enough coconut water to make a smooth pliable dough; you may not need all the coconut water. Knead the dough lightly, then cover it with a clean dish towel and set aside.

3 Make the curry. Heat the oil in a large pan, add the onion, garlic and garam masala and fry on medium heat for a few minutes to release the spice flavours. Add the cauliflower and carrot and fry for 3–4 minutes to soften.

4 Add the quinoa, lentils, coconut water, boiling water, chopped tomatoes and tomato purée to the pan and bring to the boil. Reduce the heat and simmer for 15 minutes until the quinoa and lentils are cooked. Stir in the spinach, coriander and lime pickle and cook for 2 minutes more.

5 Meanwhile, halve the flatbread dough and roll it out very thinly (about 5mm/¼in in thickness). Heat a little coconut oil in a frying pan, and fry one of the flatbreads on medium heat for 3–4 minutes, until it blisters and starts to singe slightly.

6 Flip over the bread with a spatula and cook it on the other side for 2–3 minutes, until it is starting to singe on that side too. Remove the bread from the pan, cut it into wedges or strips, then fold it in a clean dish towel to keep it warm and soft. Repeat with the remaining dough.

7 Check the curry for seasoning, then divide between four bowls, garnish with coriander and serve with lime wedges, flatbread and some lime pickle.

NUTRITIONAL INFORMATION: Energy 845kcal/3575kJ; Protein 43.8g; Carbohydrate 132.8g, of which sugars 11.8g; Fat 19.1g, of which saturates 8.5g; Cholesterol 0mg; Calcium 235mg; Fibre 15.1g; Sodium 636mg.

MARINATED SASHIMI-STYLE COCONUT TUNA

These little cubes of tuna are marinated for just 5 minutes in a delicious Japanese-style mixture that partially 'cooks' the fish. Use the freshest tuna possible.

150g/5oz very fresh tuna, skinned
10ml/2 tsp wasabi paste
30ml/2 tbsp coconut aminos or soy sauce
4 spring onions (scallions), green part only, finely chopped
50g/2oz freshly shredded coconut (see page 14), to garnish

Serves 2

1 Cut the tuna into neat 2cm/¾in cubes using a very sharp knife. Just 5–10 minutes before serving, blend the wasabi paste with the coconut aminos or soy sauce in a bowl, then add the tuna and spring onions.

2 Mix and leave to marinate for 5 minutes. Divide among two plates and garnish with shredded coconut. Serve immediately.

NUTRITIONAL INFORMATION: Energy 198kcal/825kJ; Protein 19.2g; Carbohydrate 2.2g, of which sugars 2.1g; Fat 12.6g, of which saturates 8.8g; Cholesterol 21mg; Calcium 25mg; Fibre 2.9g; Sodium 575mg.

CRAB IN COCONUT VINEGAR

White crab meat is a rich and very satisfying protein that marries well with coconut flavours in this tangy Asian-style seafood salad.

½ red (bell) pepper, seeded
salt
115g/4oz cooked white crab meat
10ml/2 tsp coconut vinegar
2.5ml/½ tsp coconut sugar
5ml/1 tsp coconut aminos or soy sauce
150g/5oz cucumber

Serves 2

1 Slice the red pepper into thin strips. Sprinkle with a little salt and leave for about 15 minutes to soften slightly. Rinse and drain.

2 Loosen the crab meat and mix it with the red pepper in a bowl. Cover with clear film (plastic wrap) and chill.

3 Combine the coconut vinegar, coconut sugar and coconut aminos or soy sauce in a bowl.

4 Cut the cucumber in half lengthways and scoop out the seeds. Finely grate the cucumber with a fine-toothed grater. Drain in a fine sieve (strainer). Mix the cucumber with the vinegar mixture.

5 Divide the crab meat mixture between two bowls, then pile half of the dressed cucumber in the centre of each. Serve immediately, before the cucumber loses its colour.

NUTRITIONAL INFORMATION: Energy 101kcal/420kJ; Protein 12.3g; Carbohydrate 5.4g, of which sugars 5.2g; Fat 3.4g, of which saturates 0.5g; Cholesterol 41mg; Calcium 18mg; Fibre 1.5g; Sodium 424mg.

COCONUT CEVICHE WITH TOMATO SALSA

You can use almost any firm-fleshed fish for this South American dish, provided that it is perfectly fresh. Coconut water adds a sweet note that complements the fish.

225g/8oz fresh sea bass fillets, skinned and cut into strips
juice of 1½ limes
60ml/4 tbsp coconut water
salt
1 fresh red chilli, finely chopped

FOR THE SALSA
75g/2oz peeled avocado
2 medium firm tomatoes, peeled, seeded and diced
15ml/1 tbsp coconut vinegar
7.5ml/1½ tsp coconut oil
15ml/1 tbsp fresh coriander (cilantro) leaves

Serves 4

1 Lay the fish in a shallow dish and pour over the lime juice and coconut water, turning to coat them all over in the juice. Cover with clear film (plastic wrap) and chill for 1 hour.

2 Season the fish with salt and sprinkle over the chilli. Toss the fish, then re-cover. Leave to marinate in the refrigerator for 15–30 minutes more.

3 For the salsa, finely dice the avocado and combine in a large bowl with the tomatoes, coconut vinegar and coconut oil.

4 Divide the salsa among four plates. Spoon on the ceviche, sprinkle the salsa with fresh coriander and serve.

NUTRITIONAL INFORMATION: Energy 108kcal/450kJ; Protein 11.8g; Carbohydrate 1.8g, of which sugars 2.4g; Fat 6g, of which saturates 1.7g; Cholesterol 45mg; Calcium 78mg; Fibre 1.9g; Sodium 82mg.

BAKED COCONUT MONKFISH WITH POTATOES

Monkfish is a firm white fish, here flavoured with fresh herbs, moistened with wine and topped with crisp crumbs. Coconut-oil roast potatoes make the perfect partner.

1kg/2¼lb waxy potatoes, cut into even chunks
75ml/5 tbsp coconut oil
2 onions, thickly sliced
4 garlic cloves
a few fresh thyme sprigs
2–3 fresh bay leaves
250ml/8fl oz/1 cup coconut water plus 45ml/3 tbsp
200ml/7fl oz/scant 1 cup vegetable or fish stock
900g/2lb monkfish tail in one piece, skin and membrane removed
30–45ml/2–3 tbsp white wine
50g/2oz/1 cup fresh white breadcrumbs
15g/½oz fresh parsley, chopped
salt and ground black pepper

Serves 4

Cook's tip
Serve the fish with some coconut mayonnaise (see page 26), stirring in 15ml/ 1 tbsp drained and chopped capers and a squeeze of fresh lime or lemon juice.

1 Preheat the oven to 190°C/375°F/Gas 5. Put the potatoes in an ovenproof dish. Melt 30ml/2 tbsp of the oil in a frying pan, add the onions and cook gently for 5–6 minutes. Add to the potatoes.

2 Slice 2–3 of the garlic cloves and add to the potatoes with the thyme and bay leaves, and season. Pour the main batch of coconut water and all of the stock over the potatoes and bake, stirring twice, for 50–60 minutes, until the potatoes are tender.

3 Nestle the monkfish into the potatoes and season. Bake for 10–15 minutes. Mix the 45ml/3 tbsp coconut water with the wine and use to baste the monkfish twice during cooking.

4 Finely chop the remaining garlic. Melt 30ml/2 tbsp of the oil and toss it with the breadcrumbs, chopped garlic, most of the chopped parsley and the seasoning. Spoon over the monkfish, pressing it down gently with the back of a spoon.

5 Drizzle the remaining 15ml/1 tbsp oil over the fish and bake for 10–15 minutes, until the breadcrumbs are crisp and golden and all the liquid has been absorbed. Sprinkle over the parsley.

NUTRITIONAL INFORMATION: Energy 537kcal/2261kJ; Protein 46g; Carbohydrate 54.5g, of which sugars 9.3g; Fat 15.9g, of which saturates 12.4g; Cholesterol 119mg; Calcium 82mg; Fibre 5.7g; Sodium 142mg.

COCONUT HAKE WITH LEMON AND CHILLI

Here coconut oil is infused with zesty lemon and spicy chilli, then brushed over hake fillets. The fish is mainly grilled skin-side up, so the flesh remains moist and tender.

30ml/2 tbsp coconut oil
2.5ml/½ tsp crushed chilli flakes
finely grated rind and juice of
   1 unwaxed lemon
4 hake fillets, each weighing about
   150g/5oz
salt and ground black pepper,
   to taste

Serves 4

Cook's tip
This simple cooking method is ideal for any thick white fish fillets, such as cod and haddock, and is also suitable for robust oily fish such as salmon fillets.

1 Put the coconut oil in a small pan. Add the chilli flakes and lemon rind and gently warm for 1 minute. Turn off the heat and brush over the hake fillets. Preheat the grill (broiler) to high.

2 Brush the fillets all over again with the oil and place skin-side up on a baking sheet. Grill (broil) the fish for 4–5 minutes, until the skin is crispy, then carefully turn them with a spatula.

3 Brush again with the coconut oil, spreading out the chilli and lemon rind evenly over the fish. Season to taste.

4 Grill the fillets for a further 2–3 minutes, or until opaque and cooked through. (Test using the point of a sharp knife; the flesh should flake.) Squeeze over the lemon juice just before serving.

NUTRITIONAL INFORMATION: Energy 64kcal/265kJ; Protein 2.8g; Carbohydrate 0.2g, of which sugars 0.2g; Fat 5.8g, of which saturates 4.8g; Cholesterol 3mg; Calcium 3mg; Fibre 0g; Sodium 15mg.

FISH IN SPICED COCONUT MILK

Both fish and coconut are found in abundance in and around the southern Indian coastline, so there are many dishes like this one, known as 'Meen Molee'. Tilapia is a white firm-textured fish or you can use monkfish, sole or cod.

675g/1½lb fillet of tilapia
30ml/2 tbsp lemon juice
5ml/1 tsp salt, or to taste
15ml/1 tbsp coconut oil
1 large onion, finely chopped
10ml/2 tsp ginger purée
5ml/1 tsp garlic purée
2 fresh green chillies, finely chopped (seeded if preferred)
2.5ml/½ tsp chilli powder
2.5ml/½ tsp ground turmeric
400ml/14oz/1⅔ cups coconut milk
45ml/3 tbsp creamed coconut, chopped, or coconut cream
cooked basmati or long grain rice, to serve (see Cook's tip)

Serves 4

Cook's tip
Prepare basmati rice by the absorption method using part stock and part coconut water for maximum flavour. Measure the rice in a measuring jug (cup) and place in a large pan with a tight-fitting lid. Measure twice the volume of liquid and add to the pan. Bring to a simmer with the lid off, then put the lid on and simmer for 8 minutes. Turn off the heat and leave the rice to stand undisturbed for 10 minutes to complete the cooking process. Do not lift the lid during this time as this will affect the cooking.

1 Cut the fish into 5cm/2in pieces and lay them out on a large plate. Rub in half the lemon juice and half the salt. Set aside.

2 Heat the oil over a medium heat, then add the onion and fry for 5–6 minutes, until the onion is soft, but not brown. Add the ginger, garlic and green chillies, and cook for 5–6 minutes, until light golden.

3 Add the chilli powder and turmeric, cook for 30 seconds, then pour in the coconut milk and add the creamed coconut or coconut cream. Add the remaining salt and lemon juice, and stir until the creamed coconut has dissolved.

4 Add the fish. Simmer for 7 minutes or until the sauce has thickened. Serve with cooked rice.

NUTRITIONAL INFORMATION: Energy 272kcal/1138kJ; Protein 31.5g; Carbohydrate 6.7g, of which sugars 6.2g; Fat 13.4g, of which saturates 9.8g; Cholesterol 0mg; Calcium 237mg; Fibre 0.2g; Sodium 203mg.

CRAB AND TOFU STIR-FRY WITH COCONUT

This simple light meal serves two, but you can easily double the quantity if needed. Canned crab meat is a useful standby ingredient, or you can use fresh white crab meat if you prefer. Serve the stir-fry with fine egg noodles or coconut rice (page 88).

200g/7oz silken tofu
30ml/2 tbsp coconut oil
2 garlic cloves, finely chopped
100g/3¾oz baby corn, halved lengthways
2 spring onions (scallions), chopped
1 fresh chilli, seeded and chopped
115g/4oz canned crab meat
30ml/2 tbsp coconut aminos or soy sauce
15ml/1 tbsp Thai fish sauce
5ml/1 tsp coconut sugar or light muscovado (brown) sugar
juice of 1 lime
30ml/2 tbsp coconut water
lime wedges and fine egg noodles or coconut rice (see page 88), to serve

Serves 2

Cook's tip
Silken tofu, also known as soft or Japanese-style tofu, has a softer consistency than regular tofu and can fall apart when cooked, so be gentle.

1 Drain the tofu and cut it into 1cm/½in cubes. Heat the coconut oil in a wok or large, heavy frying pan. Add the tofu and gently stir-fry until golden. Remove with a slotted spoon, leaving as much oil as possible behind in the pan.

2 Add the garlic to the wok or pan and stir-fry for a few seconds until beginning to colour. Add the corn, spring onions and chilli and stir-fry for 1–2 minutes until the vegetables are almost tender.

3 Return the tofu to the pan, then add the crab meat, coconut aminos or soy sauce, fish sauce, sugar, lime and coconut water.

4 Stir-fry for 1 minute until the crab and tofu are heated through and the sauce thickened slightly. Serve with lime wedges and noodles or rice.

NUTRITIONAL INFORMATION: Energy 273kcal/1145kJ; Protein 10g; Carbohydrate 40g, of which sugars 15g; Fat 8g, of which saturates 4g; Cholesterol 19mg; Calcium 209mg; Fibre 4g; Sodium 70mg.

SPICY COCONUT PAELLA

Recipes vary from region to region for this famous Spanish rice dish. This version made with coconut water balances chicken and seafood with plenty of vegetables.

2 large skinless, boneless chicken breast fillets, cut into bitesize pieces
150g/5oz prepared squid, cut into rings
8–10 raw king prawns (jumbo shrimp), shelled
325g/11oz cod or haddock fillets, skinned and cut into bitesize pieces
8 scallops, trimmed and halved
350g/12oz raw mussels in shells, scrubbed and discarding any that do not open when tapped
30ml/2 tbsp coconut oil
a bunch of spring onions (scallions), cut into strips
2 small courgettes (zucchini), cut into strips
1 red (bell) pepper, cut into strips
250g/9oz/1⅓ cups long grain rice, rinsed
150ml/¼ pint/⅔ cup chicken stock
250ml/8fl oz/1 cup coconut water
250ml/8fl oz/1 cup passata (bottled strained tomatoes)
ground black pepper
coriander (cilantro), lemon wedges, to garnish

FOR THE MARINADE
2 fresh red chillies, seeded
a handful fresh coriander (cilantro)
10ml/2 tsp ground cumin
15ml/1 tbsp paprika
2 garlic cloves
60ml/4 tbsp coconut oil, melted
juice of 1 lemon

Serves 6

1 Blend all the ingredients for the marinade in a food processor. Put the chicken pieces in one bowl and the fish and shellfish (apart from the mussels) in a separate bowl. Divide the marinade between them, mix, cover and marinate for 1 hour.

2 Drain the chicken and fish, and reserve the marinade. Heat the oil in a pan and fry the chicken until lightly browned. Add the spring onions, fry for 1 minute and then add the courgettes and pepper and fry for 3–4 minutes until slightly softened. Remove everything with a slotted spoon to separate plates.

3 Scrape the marinade into the pan and cook for 1 minute. Add the rice to the pan and stir-fry for 1 minute. Add the chicken stock, coconut water, passata and reserved chicken and stir well. Bring to the boil, then reduce the heat, cover and simmer gently for 15–20 minutes until the rice is almost tender.

4 Add the reserved vegetables to the pan and place all the fish and mussels on top. Cover and cook gently for 10–12 minutes until the fish is cooked and the mussels have opened. Discard any mussels that have not opened during cooking. Serve garnished with coriander and lemon wedges.

NUTRITIONAL INFORMATION: Energy 411kcal/1728kJ; Protein 46.5g; Carbohydrate 40g, of which sugars 6.9g; Fat 7.2g, of which saturates 3.9g; Cholesterol 212mg; Calcium 99mg; Fibre 3.1g; Sodium 575mg.

CHICKEN BREASTS WITH HAM AND COCONUT

Perfect for summer entertaining, serve this quick, light dish with new potatoes tossed in coconut oil and steamed vegetables or a salad.

4 skinless, boneless chicken breast fillets
4 slices Serrano ham
60ml/4 tbsp coconut oil
30ml/2 tbsp chopped capers
30ml/2 tbsp fresh thyme leaves
1 large lemon, cut lengthways into 8 slices
a few small fresh thyme sprigs
salt and ground black pepper
boiled new potatoes tossed in coconut oil and steamed vegetables or a tomato, (bell) pepper and olive salad, to serve

Serves 4

Cook's tip
In order to be able to shape the oil and herb mixture into four portions, you may need to chill the coconut oil in the refrigerator for 30 minutes beforehand to ensure it is hard enough, before mixing it with the capers, thyme and seasoning.

1 Preheat the oven to 200°C/400°F/Gas 6. Wrap each chicken breast loosely in clear film (plastic wrap) and beat with a rolling pin until slightly flattened. Arrange the chicken breast fillets in a large, shallow ovenproof dish, then top each with a slice of Serrano ham.

2 Mix the coconut oil with the capers, thyme and seasoning. Divide into quarters and shape each into a neat portion, then place on each ham-topped chicken breast.

3 Arrange 2 lemon slices on the coconut oil and sprinkle with small thyme sprigs. Bake for 25 minutes, or until the chicken is cooked through.

4 To serve, transfer the chicken portions to a warmed serving platter or individual plates and spoon the piquant coconut juices over the top.

5 Serve immediately, with boiled new potatoes tossed in coconut oil and steamed vegetables or a tomato salad. Discard the lemon slices before serving, if you prefer.

NUTRITIONAL INFORMATION: Energy 307kcal/1283kJ; Protein 41.1g; Carbohydrate 1.5g, of which sugars 0g; Fat 15.1g, of which saturates 10.8g; Cholesterol 105mg; Calcium 71mg; Fibre 0g; Sodium 442mg.

CHICKEN, SPLIT PEA AND COCONUT KORESH

Koresh is traditionally a hearty Persian dish, often made with lamb. Here, it has been transformed into a lighter version with coconut water, chicken and extra vegetables.

50g/2oz/¼ cup green split peas
60ml/4 tbsp coconut oil
1 large or 2 small onions, peeled and finely chopped
500g/1¼lb boneless chicken thighs
300ml/½ pint/1¼ cups chicken stock
200ml/7fl oz/scant 1 cup coconut water
5ml/1 tsp ground turmeric
2.5ml/½ tsp ground cinnamon
1.25ml/¼ tsp grated nutmeg
2 aubergines (eggplants), roughly diced
8–10 ripe tomatoes, diced
2 garlic cloves, crushed
45ml/3 tbsp chopped fresh mint, plus whole leaves to garnish
salt and ground black pepper
boiled rice, to serve

Serves 4

Variation
You could use a 400g/14oz can of chopped tomatoes instead of fresh ones. Measure and use the tomato juice instead of some of the stock.

1 Put the split peas in a bowl, pour over cold water to cover, then leave to soak for about 4 hours. Drain well.

2 Heat 15ml/1 tbsp coconut oil in a pan, add two-thirds of the onions and cook for about 5 minutes. Add the chicken and cook until golden brown on all sides.

3 Add the soaked split peas to the chicken mixture, then the stock, coconut water, turmeric, cinnamon and nutmeg. Cook over a low heat for about 40 minutes, until the peas are tender.

4 Heat the remaining coconut oil in a pan, add the aubergines and remaining onions and cook until lightly browned. Add the tomatoes, garlic and mint. Season.

5 Just before serving, stir the aubergine mixture into the chicken and split pea stew. Garnish with fresh mint leaves and serve with boiled rice.

NUTRITIONAL INFORMATION: Energy 368kcal/1542kJ; Protein 34.3g; Carbohydrate 22.1g, of which sugars 14.9g; Fat 16.5g, of which saturates 10.9g; Cholesterol 131mg; Calcium 91mg; Fibre 9.2g; Sodium 264mg.

COCONUT CHICKEN BIRYANI

In this dish the curry and rice are cooked separately, using coconut water, then layered and finished in the oven; ideal for a family supper or a relaxed dinner party.

10 whole green cardamom pods
275g/10oz/1½ cups basmati rice, soaked and drained
2.5ml/½ tsp salt
2–3 whole cloves
5cm/2in cinnamon stick
45ml/3 tbsp coconut oil
3 onions, sliced
4 boneless chicken thighs
1.25ml/¼ tsp ground cloves
1.25ml/¼ tsp hot chilli powder
5ml/1 tsp ground cumin
5ml/1 tsp ground coriander
2.5ml/½ tsp ground black pepper
3 garlic cloves, chopped
5ml/1 tsp chopped fresh root ginger
juice of 1 lemon
4 tomatoes, sliced
30ml/2 tbsp chopped fresh coriander (cilantro)
150ml/¼ pint/⅔ cup natural (plain) yogurt, plus extra to serve
4–5 saffron threads, soaked in 10ml/2 tsp hot milk
150ml/¼ pint/⅔ cup coconut water
toasted flaked (sliced) almonds and fresh coriander sprigs, to garnish

Serves 4

Cook's tip
For cucumber raita, coarsely grate half a cucumber. Place in a strainer, sprinkle with a pinch of salt and leave for 15 minutes. Squeeze out the juices and mix with 150ml/¼ pint/⅔ cup thick natural (plain) yogurt and 45ml/3 tbsp coconut cream.

1 Preheat the oven to 190°C/375°F/Gas 5. Remove the seeds from half the cardamom pods and grind them finely. Set aside.

2 Bring a pan of water to the boil and add the rice, salt, whole cardamom pods, cloves and cinnamon stick. Boil for 2 minutes, then drain, leaving the whole spices in the rice.

3 Heat the coconut oil in a frying pan and cook the onions for 8 minutes, until softened and browned. Add the chicken and the ground cloves, chilli powder, ground cumin, coriander, black pepper and the ground cardamom seeds. Mix well, then add the garlic, ginger and lemon juice. Stir-fry for 5 minutes.

4 Transfer the chicken mixture to a casserole and arrange the tomatoes on top. Top with the coriander, then the yogurt and finally the drained rice. Drizzle the saffron milk over the rice and pour over the coconut water. Cover tightly and bake for 1 hour.

5 Transfer to a warmed serving platter and remove the whole spices from the rice. Garnish with toasted almonds and fresh coriander sprigs and serve with extra yogurt or raita.

NUTRITIONAL INFORMATION: Energy 562kcal/2349kJ; Protein 33.4g; Carbohydrate 80.3g, of which sugars 22g; Fat 13.6g, of which saturates 8.4g; Cholesterol 106mg; Calcium 212mg; Fibre 6.9g; Sodium 236mg.

STIR-FRIED DUCK WITH NOODLES AND COCONUT

Removing the skin and fat from duck breasts lowers the fat content considerably and duck breast only contains a little more fat than chicken breast. Stir-fried vegetables cooked in coconut oil and noodles make this a meal in itself, perfect for entertaining.

250g/9oz fresh sesame noodles
30ml/2 tbsp coconut oil
2 duck breasts, skin removed and thinly sliced
3 spring onions (scallions), cut into strips
2 celery sticks, cut into strips
1 fresh pineapple, peeled, cored and cut into strips
300g/11oz mixed vegetables, such as carrots, (bell) peppers, beansprouts and cabbage, shredded or cut into strips
90ml/6 tbsp plum sauce

Serves 4

Cook's tip
Fresh sesame noodles can be bought from supermarkets or Asian stores. If they aren't available, use fresh egg noodles instead. Cook according to the instructions on the packet. For extra flavour, add a little coconut oil to the cooking water.

1 Cook the noodles in a pan of boiling water for 3 minutes, or according to the packet instructions. Drain in a colander.

2 Heat the oil in a wok. Add the duck and stir-fry for 2 minutes, until lightly browned. Lift out the duck with a slotted spoon and transfer to a plate, leaving behind as much oil as possible.

3 Add the spring onions and celery to the wok and stir-fry for 2 minutes more. Use a draining spoon to remove the ingredients from the wok and set aside. Add the pineapple strips and mixed vegetables, and stir-fry for 2 minutes.

4 Add the cooked noodles and plum sauce to the wok, then replace the duck, spring onion and celery mixture. Stir-fry the duck for 2 minutes more, or until the noodles and vegetables are hot and the duck is cooked through. Serve immediately.

NUTRITIONAL INFORMATION: Energy 509kcal/2151kJ; Protein 22.7g; Carbohydrate 76.7g, of which sugars 30g; Fat 14.6g, of which saturates 6.1g; Cholesterol 69mg; Calcium 101mg; Fibre 8.9g; Sodium 661mg.

SWEET AND SOUR COCONUT PORK WITH VEGETABLES

The fresh flavours, attractive colours and crisp textures of the vegetables and fruit cooked together with tender pork in coconut oil in this dish are a treat for the eyes and the stomach. Serve with some plain egg noodles or coconut rice (page 88).

350g/12oz lean pork
30ml/2 tbsp coconut oil
4 garlic cloves, thinly sliced
1 small red onion, sliced
30ml/2 tbsp Thai fish sauce
15ml/1 tbsp coconut sugar
ground black pepper, to taste
1 red (bell) pepper, seeded and diced
½ cucumber, seeded and very thinly sliced
2 plum tomatoes, cut into wedges
115g/4oz piece fresh pineapple, cut into small chunks
2 spring onions (scallions), cut into short lengths
fresh coriander (cilantro) leaves, to garnish

Serves 4

Variation
This dish is also delicious made with lean rump (round) steak. Fry for just 1 minute, as it will need less cooking time than pork.

1 Cut the pork into very thin strips; this is easier to do if you freeze it for 30 minutes first.

2 Heat the oil in a wok or large frying pan. Add the garlic. Cook over a medium heat until golden, then add the pork and stir-fry for 4–5 minutes. Add the onion slices and toss to mix.

3 Add the fish sauce, sugar and ground black pepper to taste. Toss the mixture over the heat for 3–4 minutes more.

4 Stir in the red pepper, cucumber, tomatoes, pineapple and spring onions. Stir-fry for 3–4 minutes more, then spoon into individual serving bowls. Garnish with the fresh coriander and serve immediately.

NUTRITIONAL INFORMATION: Energy 219kcal/920kJ; Protein 20.7g; Carbohydrate 14g, of which sugars 12.9g; Fat 9.4g, of which saturates 6g; Cholesterol 55mg; Calcium 36mg; Fibre 2.5g; Sodium 594mg.

STUFFED PORK WITH COCONUT

The sweet flavour and rich texture of dried fruit, such as prunes, goes particularly well with pork and coconut water and oil. If you want to ring the changes, dried apricots or figs soaked in apple juice can be used instead, with chopped walnuts.

30ml/2 tbsp coconut oil
1 shallot, very finely chopped
1 stick celery, very finely chopped
finely grated rind of ½ orange
115g/4oz/½ cup ready-to-eat pitted prunes, chopped
25g/1oz/½ cup fresh breadcrumbs
30ml/2 tbsp chopped fresh parsley
a pinch of grated nutmeg
two 225g/8oz pork fillets (tenderloins), trimmed
6 slices Parma ham or prosciutto
75ml/5 tbsp dry white wine
75ml/5 tbsp coconut water
salt and ground black pepper
mashed root vegetables and wilted pak choi (bok choy), to serve

Serves 4

Cook's tip
To turn the juices into a sauce, pour them into a frying pan and bubble them for 3–4 minutes until reduced. Stir in a little coconut cream and gently heat through. Adjust the seasoning to taste.

1 Preheat the oven to 180°C/350°F/Gas 4. Heat 15ml/1 tbsp of the coconut oil in a pan, add the shallot and celery, and fry until soft. Tip into a bowl and stir in the orange rind, prunes, breadcrumbs, parsley, nutmeg and seasoning. Leave to cool.

2 Trim off any visible fat from the pork. Slice down the length of each fillet, cutting three-quarters of the way through. Open out each pork fillet and lay it out on a board. Cover with a piece of oiled clear film (plastic wrap), then bash with a rolling pin until the meat is about 5mm/¼in thick.

3 Arrange three slices of the ham on a board and place one pork fillet on top. Repeat with the remaining ham and pork fillet. Divide the prune and breadcrumb stuffing between the pork fillets, then fold over to enclose the filling.

4 Wrap the ham around one stuffed pork fillet, and secure with one or two wooden cocktail sticks (toothpicks). Repeat with the remaining ham and pork fillet.

5 Heat the remaining coconut oil in the clean frying pan and brown the wrapped pork fillets all over, taking care not to dislodge the cocktail sticks, before transferring them into a baking dish.

6 Pour the white wine and coconut water into the frying pan and bring almost to the boil, then pour over the pork. Cover the dish with a lid or a tight covering of foil and cook in the preheated oven for 35–40 minutes or until the pork is completely cooked through and tender.

7 Allow the meat to rest somewhere warm for 5 minutes, then remove the cocktail sticks and cut into slices. Arrange on warmed plates and spoon over the cooking juices. Serve with mashed root vegetables and wilted pak choi leaves.

NUTRITIONAL INFORMATION: Energy 324kcal/1360kJ; Protein 33.5g; Carbohydrate 15.1g, of which sugars 11.5g; Fat 13.5g, of which saturates 7.4g; Cholesterol 71mg; Calcium 49mg; Fibre 3.7g; Sodium 681mg.

ADOBO CHICKEN AND PORK WITH COCONUT

Originally from Mexico, adobo has become the national dish of the Philippines. It can be made with chicken (adobong manok), with pork (adobong baboy), or with both.

30ml/2 tbsp coconut oil
6–8 garlic cloves, crushed whole
50g/2oz fresh root ginger, sliced into matchsticks
6 spring onions (scallions), cut into 2.5cm/1in pieces
5–10ml/1–2 tsp black peppercorns, crushed
30ml/2 tbsp coconut sugar
8–10 chicken thighs, or a mixture of thighs and drumsticks
350g/12oz pork fillet (tenderloin), cut into chunks
150ml/¼ pint/⅔ cup coconut vinegar
30ml/2 tbsp coconut aminos or dark soy sauce
150ml/¼ pint/⅔ cup chicken stock
150ml/¼ pint/⅔ cup coconut water
2–3 bay leaves
salt
steamed rice and stir-fried greens, to serve

Serves 4–6

Cook's tip
Coconut vinegar, or 'suka' as it is known in the Philippines, is a popular ingredient and generous amounts are added to many dishes. Once used as a way of prolonging the life of food, its sourness or 'asim' is a feature of the country's cuisine.

1 Heat the oil in a wok with a lid or in a large, heavy pan. Stir in the garlic and ginger and fry until they become fragrant and begin to colour. Add the spring onions and black peppercorns and stir in the sugar.

2 Add the chicken and pork to the wok or pan and fry until they begin to colour.

3 Pour in the vinegar, coconut aminos or soy sauce, chicken stock and coconut water and add the bay leaves. Bring to the boil, then reduce the heat, cover and simmer gently for about 1 hour, until the meat is tender and the liquid has reduced.

4 Season the stew with salt to taste and serve with stir-fried greens and rice, over which the cooking liquid is spooned.

NUTRITIONAL INFORMATION: Energy 397kcal/1663kJ; Protein 50.1g; Carbohydrate 7.3g, of which sugars 8.2g; Fat 18.8g, of which saturates 7.6g; Cholesterol 234mg; Calcium 35mg; Fibre 1g; Sodium 617mg.

SPICED COCONUT LAMB WITH TOMATOES AND PEPPERS

Select lean and tender leg of lamb for this lightly spiced curry, which is made with coconut oil, tomatoes and succulent chunks of pepper and wedges of onion.

1.5kg/3¼lb lean boneless lamb, cubed
250ml/8fl oz/1 cup natural (plain) yogurt
30ml/2 tbsp coconut oil
3 onions
2 red (bell) peppers, seeded and cut into chunks
3 garlic cloves, finely chopped
1 fresh red chilli, seeded and chopped
2.5cm/1in piece fresh root ginger, peeled and chopped
30ml/2 tbsp mild curry paste
2 x 400g/14oz cans chopped tomatoes
salt and ground black pepper
a large pinch of saffron strands
800g/1¾lb plum tomatoes, halved, seeded and cut into chunks
chopped fresh coriander (cilantro), to garnish
warm naan, roti or chapattis, to serve

Serves 6

1 Mix the lamb with the yogurt in a bowl. Cover and chill for about 1 hour. Heat the oil in a large pan. Drain the lamb and reserve the yogurt, then cook the lamb in batches until it is golden on all sides. Remove from the pan and set aside.

2 Cut two of the onions into wedges and add to the oil remaining in the pan. Fry for about 10 minutes, until they begin to colour. Add the peppers and cook for a further 5 minutes. Remove the vegetables from the pan and set aside.

3 Chop the remaining onion. Add it to the pan with the garlic, chilli and ginger, and cook, stirring, until softened. Stir in the curry paste and tomatoes with the reserved marinade. Replace the lamb, add seasoning to taste and stir. Bring to the boil, reduce the heat to low and simmer for 30 minutes.

4 Pound the saffron to a powder in a mortar, then stir in a little boiling water to dissolve it. Add to the curry. Replace the onion and pepper mixture. Stir in the fresh tomatoes and bring back to simmering point, then cook for 15 minutes. Garnish with chopped fresh coriander and serve with warm bread.

NUTRITIONAL INFORMATION: Energy 587kcal/2454kJ; Protein 57.5g; Carbohydrate 24.2g, of which sugars 22.5g; Fat 29.6g, of which saturates 14.5g; Cholesterol 194mg; Calcium 164mg; Fibre 7.1g; Sodium 435mg.

MADRAS BEEF CURRY WITH COCONUT RICE

Chillies are an indispensable ingredient of hot and spicy Madras curry. After long, gentle simmering, their flavour mellows and they merge with the other flavourings to give a delectable result that goes perfectly with coconut and vegetable rice.

60ml/4 tbsp coconut oil
675g/1½lb stewing beef, cut into bitesize cubes
1 onion, chopped
3 green cardamom pods
2 fresh green chillies, seeded and finely chopped
2.5cm/1in piece of fresh root ginger, grated
2 garlic cloves, crushed
15ml/1 tbsp Madras curry paste
10ml/2 tsp ground cumin
7.5ml/1½ tsp ground coriander
75ml/5 tbsp well-flavoured beef stock
75ml/5 tbsp coconut water
salt
coconut cream or natural (plain) yogurt, to serve (optional)

FOR THE COCONUT RICE
225g/8oz/generous 1 cup basmati rice
15ml/1 tbsp coconut oil
25g/1oz/2 tbsp butter
1 onion, finely chopped
1 garlic clove, crushed
5ml/1 tsp ground cumin
2.5ml/½ tsp ground coriander
4 green cardamom pods
1 cinnamon stick
1 each small red and green (bell) pepper, seeded and diced
150ml/¼ pint/⅔ cup vegetable stock
150ml/¼ pint/⅔ cup coconut water

Serves 4

1 Heat half the coconut oil in a large, shallow pan. When it is hot, fry the meat, in batches if necessary, until it is browned on all sides. Transfer to a plate and set aside.

2 Heat the remaining coconut oil in the pan and fry the onion for 3–4 minutes until it is softened and lightly browned. Add the cardamom pods and fry for 1 minute, then stir in the chillies, ginger and garlic, and fry for 2 minutes more.

3 Stir in the curry paste, ground cumin and ground coriander, then return the meat to the pan. Stir in the stock and coconut water. Season, bring to the boil, then reduce the heat and simmer very gently for 1–1½ hours, until the meat is tender.

4 When the curry is almost ready, prepare the spicy rice. Put the basmati in a bowl and pour over enough boiling water to cover. Set aside for 10 minutes, then drain, rinse under cold water and drain again.

5 Heat the coconut oil and butter in a flameproof casserole and fry the onion and garlic gently for 4–5 minutes until softened and lightly browned.

6 Stir in the ground cumin and coriander, the cardamom pods and the cinnamon stick. Fry for 1 minute, then add the diced peppers. Add the rice, stirring well to coat the grains in the spice mixture, then pour in the stock and coconut water.

7 Bring to the boil, then reduce the heat, cover the pan tightly and simmer for about 8–10 minutes, or until the rice is tender and the liquid has been absorbed.

8 Spoon the spicy rice into a bowl and serve immediately with the curry. You can also serve this with a generous dollop of coconut cream or natural yogurt if you like.

NUTRITIONAL INFORMATION: Energy 575kcal/2401kJ; Protein 44.2g; Carbohydrate 49.9g, of which sugars 4.6g; Fat 23g, of which saturates 14.9g; Cholesterol 98mg; Calcium 82mg; Fibre 1.7g; Sodium 219mg.

DESSERTS

COCONUT AND LEMON GRASS ICE CREAM

Lemon grass adds an exotic fragrance to this smooth and creamy coconut ice cream. If you can't get the fresh ingredient, use the dried stalks or preserved stalks in jars.

4 lemon grass stalks, sliced lengthways and bruised with a rolling pin
400ml/14fl oz/1⅔ cups coconut milk
3 egg yolks
90g/3½oz/½ cup caster (superfine) sugar
10ml/2 tsp cornflour (cornstarch)
150ml/¼ pint/⅔ cup coconut cream
finely grated rind of 1 lime

FOR THE LIME SYRUP
75g/3oz/6 tbsp caster (superfine) sugar
75ml/5 tbsp coconut water
1 lime, very thinly sliced, plus 30ml/2 tbsp lime juice

Serves 5–6

Variation
You can use coconut sugar instead of caster (superfine) sugar in this ice cream if you prefer; it will give it a delicious caramel flavour and a slightly darker colour.

1 Put the lemon grass in a heavy pan, add the coconut milk and bring to just below boiling point. Remove from the heat and leave to infuse for 30 minutes, then remove the lemon grass.

2 Whisk the yolks in a bowl with the sugar and cornflour until smooth. Slowly add the coconut milk, whisking constantly. Return to the pan and heat gently, stirring until the custard thickens. Strain into a clean bowl. Cover with baking parchment and chill.

3 Stir the coconut cream and lime rind into the custard. Churn in an ice cream maker until thick, then spoon into 5–6 dariole moulds. Freeze for at least 3 hours.

4 Heat the sugar and coconut water in a pan until the sugar dissolves. Boil for 5 minutes without stirring. Reduce the heat, add the lime slices and juice and simmer for 5 minutes. Cool.

5 To turn out, loosen with a knife and briefly dip in very hot water. Serve with the syrup and lime slices.

NUTRITIONAL INFORMATION: Energy 161kcal/681kJ; Protein 2.1g; Carbohydrate 33.5g, of which sugars 32.8g; Fat 3g, of which saturates 0.9g; Cholesterol 101mg; Calcium 39mg; Fibre 0.4g; Sodium 112mg.

WATERMELON AND COCONUT ICE

This stunning pink sorbet made with coconut water is simple to make and has a refreshing fruity flavour. It is perfect for cooling down on a hot sunny day.

90ml/6 tbsp caster (superfine) sugar
105ml/7 tbsp coconut water
4 kaffir lime leaves, torn into small pieces
500g/1¼lb watermelon

Serves 4–6

Variations
- If you can't find kaffir lime leaves, use a couple of pared strips of lime zest and a generous squeeze of lime juice instead.
- To make a watermelon granita, pour the mixture into a metal baking tray with sides and freeze for 1 hour. Stir well with a fork, mashing any ice crystals, then return to the freezer for 2 hours. Using a fork, scrape the frozen mixture to create flakes of ice. Serve the granita immediately, or transfer it to a freezerproof container, where it will keep, ready to use, for up to 3 days.

1 Put the sugar, coconut water and lime leaves in a heavy pan and heat gently until the sugar has dissolved. Pour into a bowl and set aside.

2 Cut the watermelon into wedges with a large knife. Cut the flesh from the rind, remove the seeds and chop the flesh. Transfer to a food processor, process to a slush, then mix with the syrup. Chill the mixture in the refrigerator for 3–4 hours.

3 Strain the mixture into a freezerproof container. Freeze for 2 hours, then beat with a fork to break up the ice crystals. Freeze for 3 hours more, beating at half-hourly intervals. Freeze until firm.

4 Alternatively, use an ice cream maker. Pour the chilled mixture into the machine and churn until it is firm enough to scoop. Serve immediately, or freeze in a freezerproof container.

5 About 30 minutes before serving, transfer the ice to the refrigerator so that it softens slightly and can be scooped.

NUTRITIONAL INFORMATION: Energy 86kcal/369kJ; Protein 0.9g; Carbohydrate 21.6g, of which sugars 22.6g; Fat 0.3g, of which saturates 0.1g; Cholesterol 0mg; Calcium 10mg; Fibre 0.6g; Sodium 47mg.

COCONUT AND COFFEE TRIFLE

Dark coffee sponge laced with liqueur, coconut custard and a coffee-cream topping makes this a lavish dessert. Serve in a large glass bowl for maximum impact.

FOR THE COFFEE SPONGE
45ml/3 tbsp ground coffee
45ml/3 tbsp near-boiling water
2 eggs
50g/2oz/¼ cup coconut sugar or soft dark brown sugar
40g/1½oz/⅓ cup self-raising (self-rising) flour, sifted
25ml/1½ tbsp coconut oil

FOR THE COCONUT CUSTARD
400ml/14fl oz/1⅔ cup coconut milk
3 eggs
45ml/3 tbsp caster (superfine) sugar
10ml/2 tsp cornflour (cornstarch)

FOR THE FILLING AND TOPPING
2 medium bananas, sliced
60ml/4 tbsp coconut or coffee liqueur
150ml/¼ pint/⅔ cup whipping cream
30ml/2 tbsp icing (confectioners') sugar, sifted
150ml/¼ pint/⅔ cup whipped coconut cream (see page 26)
5ml/1 tsp vanilla extract
ribbons of fresh coconut, to decorate

Serves 6–8

1 Preheat the oven to 160°C/325°F/Gas 3. Grease and line an 18cm/7in square tin (pan) with baking parchment. Put the coffee in a small bowl. Pour over the hot water and leave to infuse for 4 minutes. Strain, discarding the grounds.

2 Whisk the eggs and sugar until the whisk leaves a trail when lifted. Gently fold in the flour, followed by 15ml/1 tbsp of the coffee and the oil. Spoon into the tin and bake for 20 minutes. Turn out on to a wire rack, remove the paper and leave to cool.

3 To make the custard, heat the coconut milk in a pan until it is almost boiling. Whisk the eggs, sugar and cornflour together until frothy. Pour on the milk, whisking all the time. Add to the pan and heat gently, stirring for 1–2 minutes, until the custard thickens, but do not boil. Leave to cool for 10 minutes, stirring occasionally.

4 Cut the sponge into 5cm/2in squares and arrange in the base of a large glass bowl. Arrange the banana on top, then drizzle over the liqueur. Pour the custard over and leave until cold.

5 Whip the cream with the remaining coffee and icing sugar until soft peaks form, then fold in the whipped coconut cream and vanilla extract. Spoon over the custard. Cover and chill. Sprinkle with ribbons of fresh coconut and serve.

NUTRITIONAL INFORMATION: Energy 288kcal/1204kJ; Protein 7.7g; Carbohydrate 27.4g, of which sugars 23g; Fat 15.3g, of which saturates 8.1g; Cholesterol 212mg; Calcium 90mg; Fibre 0.6g; Sodium 206mg.

Variation: Chocolate and Banana Fool
Melt 175g/6oz plain (semisweet) chocolate. Pour the coconut custard (see recipe above) into a bowl and fold in the melted chocolate to create a rippled effect. Peel and slice 3 medium bananas and stir these into the mixture. Spoon into 6 glasses and chill for at least 30 minutes before serving. Serves 6.

NUTRITIONAL INFORMATION: Energy 321kcal/1354kJ; Protein 6.5g; Carbohydrate 49.3g, of which sugars 46.4g; Fat 12.3g, of which saturates 6.3g; Cholesterol 117mg; Calcium 90mg; Fibre 1.7g; Sodium 266mg.

APRICOT AND COCONUT DESSERT

This mouthwatering Turkish dessert is traditionally made with fresh apricots, which are cooked with sugar and warm fragrant spices until soft, but this quick and easy version uses dried, ready-to-eat apricots and coconut water.

500g/1¼lb dried ready-to-eat apricots, preferably unsulphured
600ml/1 pint/2½ cups coconut water
4 cloves
2 star anise
50g/2oz/¼ cup coconut sugar or soft light brown sugar
whipped coconut cream (see page 26), to serve
30–45ml/2–3 tbsp toasted pistachio nuts, lightly crushed, to garnish

Serves 4

Variation
Mixed dried fruit such as apples, pears and peaches would also be good prepared this way. Top with chopped toasted walnuts instead of pistachio nuts.

1 Put the apricots into a pan with the coconut water and 150ml/ ¼ pint/⅔ cup water, the cloves and the star anise, and bring to the boil. Reduce the heat to low, cover the pan and simmer for 15 minutes, or until the apricots are soft. Stir at least twice during this time to ensure even cooking.

2 Remove the spices from the pan and discard them. Remove half of the apricots with a slotted spoon and set aside. Purée the remainder, along with the cooking juices, and return the purée to the pan. Add the whole apricots.

3 Add the sugar and stir to combine thoroughly. Cook gently for 3–4 minutes, then remove the pan from the heat and allow to cool for 30 minutes.

4 Divide between four serving dishes, top with some whipped coconut cream and the crushed pistachio nuts and serve.

NUTRITIONAL INFORMATION: Energy 259kcal/1104kJ; Protein 8.1g; Carbohydrate 58.7g, of which sugars 67.7g; Fat 0.8g, of which saturates 0g; Cholesterol 0mg; Calcium 95mg; Fibre 15g; Sodium 396mg.

CARAMELIZED PLUMS WITH STICKY COCONUT RICE

Red, juicy plums are quickly seared with coconut sugar to make a rich caramel coating, then served with sticky coconut-flavoured rice for a satisfying dessert. Glutinous rice is available from Asian stores; remember that it needs to be soaked overnight.

90g/3½oz/½ cup coconut sugar or caster (superfine) sugar
6 or 8 firm, ripe plums, halved and pitted

FOR THE RICE
115g/4oz sticky glutinous rice
150ml/¼ pint/⅔ cup coconut cream
45ml/3 tbsp caster (superfine) sugar
a pinch of salt

Serves 4

Cook's tip
Using coconut sugar for caramelizing the plums will give a darker, richer colour and flavour, but you can use caster (superfine) sugar if you prefer.

Variation
Other stone fruits, such as fresh ripe apricots, nectarines or peaches, would also work very well instead of plums in this recipe.

1 First prepare the rice. Rinse it in several changes of water, then leave to soak overnight in a bowl of cold water.

2 Line a large bamboo steamer with muslin (cheesecloth). Drain the rice and spread it out evenly on the muslin. Cover the rice and steam over simmering water for 25–30 minutes, until the rice is tender. (Check the water level and add more if necessary.) Transfer the rice to a wide bowl and set it aside for a moment.

3 Put the coconut cream, sugar and salt into a heavy pan. Heat gently, stirring until the sugar has dissolved, then bring to a boil. Remove from the heat and pour over the rice. Stir to mix well.

4 Sprinkle sugar over the cut sides of the halved plums. Heat a non-stick frying pan or wok over a medium-high heat. Working in batches, place the plums in the pan, cut-side down, and cook for 1–2 minutes, or until the sugar caramelizes. (You may have to wipe out the pan with kitchen paper in between batches.)

5 Mould the rice into rounds and place on warmed plates or spoon into bowls, then top with the caramelized plums.

NUTRITIONAL INFORMATION: Energy 407kcal/1712kJ; Protein 4.8g; Carbohydrate 68.7g, of which sugars 47.2g; Fat 13.6g, of which saturates 11.2g; Cholesterol 0mg; Calcium 33mg; Fibre 2.4g; Sodium 6mg.

COCONUT CUSTARD

This classic Thai dessert is made with creamy coconut milk and slowly oven-cooked. Similar to crème brûlée, it has a rich nutty flavour and simple coconut sugar topping.

4 eggs
75g/3oz/generous ⅓ cup coconut sugar or soft light brown sugar
250ml/8fl oz/1 cup coconut milk
5ml/1 tsp vanilla, rose or jasmine extract
icing (confectioners') sugar, to decorate
sliced fresh fruit, to serve

Serves 4

Variation
If liked, these can be baked in 150ml/¼ pint/⅔ cup ramekins and turned out to serve. Line each ramekin with a small circle of baking parchment and grease with a little coconut oil. After cooking, chill the custards for at least 2 hours, then loosen by running a knife around the edge, turn out on to plates and remove the parchment. Serve with a selection of tropical fruits, such as mango, papaya or tamarillos.

1 Pour about 2.5cm/1in hot water into the base of a deep baking tray and heat the oven to 160°C/325°F/Gas 3. Whisk the eggs and sugar in a bowl until smooth. Gradually add the coconut milk and flavoured extract, and whisk well.

2 Strain the mixture into a jug or pitcher, then pour into four individual heatproof glasses, ramekins or one single ovenproof dish. Cover the containers with clear film (plastic wrap).

3 Place the dishes in the baking tray. If necessary, pour more boiling water around them to reach just over halfway up their sides. Cover the baking tray with foil, then cook for 35–40 minutes, or until the custards are lightly set. Test with a fine skewer or cocktail stick (toothpick); it should come out clean.

4 Carefully lift out the dishes from the baking tray and leave to cool. Once cool, chill in the refrigerator until ready to serve. Decorate with a light dusting of icing sugar, and serve with fruit.

NUTRITIONAL INFORMATION: Energy 178kcal/751kJ; Protein 7.8g; Carbohydrate 22.7g, of which sugars 22.7g; Fat 6.9g, of which saturates 2g; Cholesterol 231mg; Calcium 58mg; Fibre 0g; Sodium 154mg.

COCONUT TAPIOCA PUDDING

Another Thai-style dessert, this is made from large pearl tapioca, coconut water and coconut milk and served warm topped with fresh lychees and shaved fresh coconut.

115g/4oz/⅔ cup large pearl tapioca
475ml/16fl oz/2 cups coconut water
115g/4oz/generous ½ cup caster (superfine) sugar
a pinch of salt
250ml/8fl oz/1 cup coconut milk
250g/9oz prepared tropical fruits
shredded lime rind and shaved fresh coconut, to decorate (optional)

Serves 4

Variations
• This dish includes a lot of sugar – as it would in Thailand – but you may prefer to reduce the sugar according to taste.
• You could use coconut sugar in place of caster (superfine) sugar if you like, although this will alter the appearance and flavour of the finished dish.

1 Put the tapioca in a bowl and pour over enough warm water to cover generously. Leave the tapioca to soak for 1 hour until the grains swell, then drain well and set aside.

2 Pour the coconut water into a heavy pan and place over a medium heat. Add the sugar and salt and stir until dissolved. Bring the mixture to the boil.

3 Add the tapioca and coconut milk and stir well. Cover and cook for a further 45–50 minutes, or until the tapioca grains become transparent.

4 Spoon into one large dish or four individual bowls and serve warm with tropical fruits, decorated with the lime rind and coconut shavings, if using.

NUTRITIONAL INFORMATION: Energy 275kcal/1176kJ; Protein 3.3g; Carbohydrate 69.2g, of which sugars 48.9g; Fat 0.4g, of which saturates 0.2g; Cholesterol 0mg; Calcium 36mg; Fibre 5.9g; Sodium 372mg.

YOUNG COCONUT PUDDING

'Podeng' (from the dish's name of 'podeng kelapa muda') is the Indonesian transliteration of 'pudding'. This one is very similar to the Western 'queen of puddings' but has a coconut rather than lemon-and-strawberry-jam flavour.

6 egg yolks
50g/2oz/½ cup caster (superfine) sugar
1.25ml/¼ tsp coconut or vanilla extract
65g/2½oz/9 tbsp plain (all-purpose) flour
90ml/6 tbsp sweetened condensed milk
120ml/4fl oz/½ cup warm coconut water
250g/9oz young coconut flesh, finely chopped

FOR THE MERINGUE
6 egg whites
15ml/1 tbsp caster (superfine) sugar
a pinch of salt
60ml/4 tbsp flaked (sliced) and toasted almonds

Serves 6

Variation
You can use toasted coconut flakes instead of almonds for the topping, if you prefer.

1 Preheat the oven to 200°C/400°F/Gas 6. Put the yolks, sugar and vanilla extract in a large bowl and whisk until slightly frothy. Gradually add the flour to the condensed milk and coconut water in a medium pan, mixing constantly to blend well.

2 Mix the coconut flesh into the flour mixture, then stir in the egg yolk and sugar mixture. Cook over a low heat, stirring with a wooden spoon for about 5 minutes, until the mixture thickens. Pour into a greased ovenproof dish.

3 Put the egg whites in a clean, grease-free bowl and whisk until white and stiffening, using an electric whisk if you prefer. Add the sugar and salt and continue to whisk until stiff peaks form.

4 Spread the meringue over the coconut mixture and bake for 15–20 minutes, until the topping turns pale golden. Top with the flaked and toasted almonds and serve at room temperature.

NUTRITIONAL INFORMATION: Energy 411kcal/1714kJ; Protein 12g; Carbohydrate 30.3g, of which sugars 23g; Fat 27.7g, of which saturates 16.1g; Cholesterol 207mg; Calcium 116mg; Fibre 5.1g; Sodium 151mg.

PINEAPPLE WITH PAPAYA AND COCONUT SAUCE

A delicious yet simple dessert, here pineapple slices are sprinkled with coconut sugar and sweet spices before being grilled (broiled). A fragrant pinky-orange papaya sauce made with coconut water complements the pineapple perfectly.

1 sweet pineapple
7.5ml/1½ tsp coconut oil, for greasing
2 pieces drained stem ginger in syrup, cut into fine matchsticks, plus 30ml/2 tbsp of the syrup
30ml/2 tbsp coconut sugar or demerara (raw) sugar
a pinch of ground cinnamon

FOR THE SAUCE
1 ripe papaya, peeled and seeded
175ml/6fl oz/¾ cup coconut water

Serves 6

Cook's tips
- You can use the papaya sauce made in this recipe as an accompaniment to savoury dishes, too. It tastes great with grilled (broiled) chicken and game birds, as well as pork and lamb.
- Make the sauce in advance and freeze it in an ice-cube tray. Defrost as needed.

1 Peel the pineapple and cut spiral slices off the outside to remove the eyes. Cut the pineapple into six 2.5cm/1in-thick slices.

2 Line a baking sheet with foil, rolling up the sides to make a rim. Grease the foil with the coconut oil.

3 Preheat the grill (broiler). Arrange the pineapple on the baking sheet. Top with the ginger matchsticks, sugar and cinnamon. Drizzle over the stem ginger syrup. Grill (broil) for 5–7 minutes, until the slices are golden and lightly charred on top.

4 Meanwhile, make the sauce. Cut a few slices from the papaya and set aside, then purée the rest with the coconut water in a food processor or blender.

5 Sieve (strain) the purée, then stir in any cooking juices from the pineapple. Serve the pineapple drizzled with the sauce and decorated with the papaya slices.

NUTRITIONAL INFORMATION: Energy 88kcal/375kJ; Protein 1.7g; Carbohydrate 20.6g, of which sugars 19.3g; Fat 0.4g, of which saturates 0g; Cholesterol 0mg; Calcium 42mg; Fibre 4g; Sodium 98mg.

COCONUT MANGO STACKS WITH RASPBERRY COULIS

Filo pastry is a healthier alternative to shortcrust or puff pastry. Brushing the sheets with coconut oil makes it beautifully crisp and golden when baked. Fill the pastry rounds with fresh mango slices and raspberries for a really special dessert.

3 filo pastry sheets, thawed if frozen
50g/2oz/¼ cup coconut oil, melted
2 small ripe mangoes
115g/4oz/⅔ cup raspberries, thawed if frozen
45ml/3 tbsp coconut water

Serves 4

Cook's tips
- Serve as soon as the mango stacks are assembled as the pastry will quickly absorb moisture from the fruit and lose its crispness.
- Fill the filo pastry layers with any fresh or cooked fruit of your choice. Poached apple slices are delicious with a bramble purée or try sliced peaches or nectarines with a strawberry purée. You could also add a spoonful or two of whipped coconut cream (see page 26) between the pastry layers.

1 Preheat the oven to 200°C/400°F/Gas 6. Lay the filo sheets on a clean work surface and cut out four 10cm/4in rounds from each.

2 Brush each round with the melted coconut oil and lay the rounds on two baking sheets. Bake for 5 minutes, or until crisp and golden. Place on wire racks to cool.

3 Peel the mangoes, remove the stones (pits) and cut the flesh into thin slices. Blend the raspberries in a food processor with the coconut water to make a purée.

4 Place a pastry round on each of four plates. Top with a quarter of the mango and drizzle with raspberry purée. Repeat until all the ingredients are used, finishing with a layer of mango and a dash of purée.

NUTRITIONAL INFORMATION: Energy 253kcal/1068kJ; Protein 4.7g; Carbohydrate 41g, of which sugars 12.9g; Fat 9g, of which saturates 7.3g; Cholesterol 0mg; Calcium 69mg; Fibre 5.5g; Sodium 32mg.

COCONUT TARTS

In the Philippines, where coconuts are plentiful, many desserts are made from young coconuts when the inside is soft and jelly-like. At this stage the flesh isn't naturally sweet and has quite a different flavour than that of mature coconuts.

FOR THE PASTRY
225g/8oz/2 cups plain (all-purpose) flour
a pinch of salt
5ml/1 tsp coconut sugar or soft light brown sugar
60ml/4 tbsp coconut oil, melted and cooled
30ml/2 tbsp warm coconut water
1 egg, at room temperature

FOR THE FILLING
6 egg yolks
75ml/5 tbsp coconut sugar or soft light brown sugar
60ml/4 tbsp plain (all-purpose) flour
45ml/3 tbsp coconut oil, melted
600g/1lb 6oz young coconut flesh, finely chopped
5ml/1 tsp lime juice

Makes 12 tarts

1 To make the pastry, mix the flour, salt and sugar in a bowl. Combine the coconut oil, water and egg. Make a well in the flour, add the oil mixture and slowly mix in the flour to make a dough.

2 Knead the dough on a lightly floured surface for a few seconds until smooth, then cover with a damp dish towel and leave to rest for 30 minutes; it will soften and become more pliable.

3 Thinly roll out the pastry on a floured surface and stamp out 12 rounds 10cm/4in in diameter. Use to line the holes of a greased deep tartlet tin or muffin pan. Chill for 30 minutes.

4 Meanwhile, place a baking sheet in the oven and preheat to 180°C/350°F/Gas 4. Beat the egg yolks and sugar until the sugar is dissolved, then mix the flour and melted coconut oil together and add to the egg yolks. Mix well. Add the coconut and lime juice, then spoon some of the mixture into each tart case.

5 Put the tartlet tin on the hot baking sheet and bake for 15–20 minutes or until the pastry is golden and the filling lightly set.

NUTRITIONAL INFORMATION: Energy 375kcal/1557kJ; Protein 6.1g; Carbohydrate 20.9g, of which sugars 2.8g; Fat 30.2g, of which saturates 24.2g; Cholesterol 120mg; Calcium 55mg; Fibre 6.5g; Sodium 22mg.

BAKING

BARBADIAN COCONUT SWEET BREAD

Often made at Christmas time in Barbados, this dark and sticky coconut cake-like bread is delicious served with a fruity rum punch or a cup of frothy hot chocolate.

225g/8oz/2 cups self-raising (self-rising) flour
150g/5oz/1¼ cups plain (all-purpose) flour
25g/1oz/¼ cup coconut flour
175g/6oz/¾ cup coconut oil, chilled for 10 minutes
115g/4oz/½ cup coconut sugar
115g/4oz/1½ cups desiccated (dry unsweetened shredded) coconut
5ml/1 tsp mixed (apple pie) spice
10ml/2 tsp vanilla extract
15ml/1 tbsp dark rum (optional)
2 eggs
about 150ml/¼ pint/⅔ cup coconut-flavoured milk or full-fat milk
15ml/1 tbsp caster (superfine) sugar, blended with 30ml/2 tbsp water, to glaze

Makes 2 small loaves or 1 large loaf

Variation
You could add dried fruit such as raisins or even dark (bittersweet) chocolate chips to the mixture before baking.

1 Preheat the oven to 180°C/350°F/Gas 4. Grease two 450g/1lb loaf tins (pans) or one 900g/2lb tin.

2 Sift the three flours into a large bowl. Add the chilled coconut oil and, using a pastry cutter or a knife, cut it into the flour until the mixture resembles breadcrumbs. Stir in the coconut sugar.

3 Add the shredded coconut, mixed spice, vanilla extract, rum, if using, eggs and milk. Mix together well with your hands. If the mixture is too dry, add more milk.

4 Transfer the dough to a lightly floured board and knead for a few minutes until it is firm and pliable.

5 Halve the mixture and place in the loaf tins. Glaze with sugared water and bake for about 1 hour, or until a skewer inserted into the loaf comes out clean.

NUTRITIONAL INFORMATION: Energy 3279kcal/13700kJ; Protein 54.5g; Carbohydrate 320.8g, of which sugars 20.3g; Fat 207g, of which saturates 175.1g; Cholesterol 0mg; Calcium 1141mg; Fibre 37.5g; Sodium 1244mg.

SULTANA, COCONUT AND WALNUT BREAD

This fruit and nut loaf tastes good plain or can be served buttered or spread with a little coconut oil and topped with coconut syrup or jam. It is also lovely toasted.

300g/11oz/2¾ cups strong white bread flour
2.5ml/½ tsp salt
15ml/1 tbsp coconut oil, chilled for 10 minutes
7.5ml/1½ tsp easy-blend (rapid-rise) dried yeast
175ml/6fl oz/¾ cup lukewarm coconut water
115g/4oz/scant 1 cup sultanas (golden raisins)
75g/3oz/½ cup walnuts or brazil nuts, roughly chopped
coconut oil, for brushing

Makes 1 loaf

Variations
• For a richer loaf, use coconut-flavoured milk instead of the coconut water.
• If liked, add 10ml/2 tsp ground cinnamon when sifting the flour.

1 Sift the flour and salt into a bowl, cut in the coconut oil with a knife, then stir in the yeast. Gradually add the coconut water to the flour mixture, stirring with a spoon at first, then gathering the dough together with your hands.

2 Turn the dough out on to a floured surface and knead for about 10 minutes until smooth and elastic. Knead the sultanas and nuts into the dough until they are evenly distributed.

3 Shape into a rough oval, place on a lightly oiled baking sheet and cover with oiled clear film (plastic wrap). Leave to rise in a warm place for 1–2 hours, until doubled in bulk. Preheat the oven to 220°C/425°F/Gas 7.

4 Uncover and bake for 10 minutes, then reduce the temperature to 190°C/375°F/Gas 5 and bake for a further 20–25 minutes.

5 Transfer to a wire rack, brush with coconut oil and cover with a dish towel. Cool before slicing.

NUTRITIONAL INFORMATION: Energy 1340kcal/5685kJ; Protein 33.3g; Carbohydrate 285.1g, of which sugars 67g; Fat 15.2g, of which saturates 10.1g; Cholesterol 0mg; Calcium 455mg; Fibre 19.7g; Sodium 1478mg.

COCONUT AND PINEAPPLE CARROT CAKE

The pineapple in this classic carrot cake recipe gives it extra moistness as well as a fabulous fruity boost. You can leave the cake plain and serve it with whipped coconut cream or spread it with the mascarpone topping as suggested in the recipe.

250g/9oz/2¼ cups plain (all-purpose) flour
10ml/2 tsp baking powder
5ml/1 tsp bicarbonate of soda (baking soda)
2.5ml/½ tsp salt
5ml/1 tsp ground cinnamon
45ml/3 tbsp poppy seeds
225g/8oz/1⅓ cups coconut sugar or soft light brown sugar
3 eggs, beaten
finely grated rind of 1 orange
225g/8oz raw carrots, grated
75g/3oz/½ cup fresh or canned pineapple, drained and finely chopped
75g/3oz/¾ cup walnut pieces
115g/4oz/½ cup coconut oil, melted

FOR THE MASCARPONE ICING
150g/5oz/scant 1 cup mascarpone
30ml/2 tbsp icing (confectioners') sugar, sifted
finely grated rind of 1 orange

Makes 1 large loaf

1 Preheat the oven to 180°C/350°F/ Gas 4. Line the base of a 1.5 litre/2½ pint/6¼ cup loaf tin (pan) with baking parchment. Grease the sides of the tin and dust with flour.

2 Sift together the flour, baking powder, bicarbonate of soda, salt and cinnamon in a bowl. Stir in the poppy seeds.

3 Mix together the sugar, eggs and orange rind in a separate bowl. Lightly squeeze the excess moisture from the carrots and stir them into the egg mixture with the pineapple and walnut pieces. Gradually stir the sifted flour mixture into the egg mixture until well combined, then gently fold in the oil.

4 Spoon into the tin, level the top and bake for 1–1¼ hours, until golden brown. Test the cake with a skewer: if it comes out clean the cake is done, if not, cook it for another 10 minutes. Remove the cake from the tin and allow it to cool on a wire rack. Remove the baking parchment when it is completely cold.

5 To make the icing, beat the mascarpone with the icing sugar and orange rind. Spread it thickly over the top of the cake.

NUTRITIONAL INFORMATION: Energy 4356kcal/18228kJ; Protein 82.7g; Carbohydrate 494.3g, of which sugars 296.4g; Fat 241g, of which saturates 132.2g; Cholesterol 765mg; Calcium 1109mg; Fibre 27g; Sodium 4972mg.

CHOCOLATE, COCONUT AND PRUNE CAKE

This cake has a rich, deep chocolate flavour and a dense, moist, brownie-like texture provided by puréed prunes and coconut flour. No extra sugar is added to the mixture as enough sweetness is provided by both the chocolate and the prunes.

200g/7oz dark (bittersweet) chocolate
200g/7oz generous 1 cup ready-to-eat pitted prunes
3 eggs, beaten
75ml/5 tbsp coconut oil, melted
115g/4oz/1 cup self-raising (self-rising) flour
25g/1oz/¼ cup coconut flour
7.5ml/1½ tsp baking powder
200ml/7fl oz/scant 1 cup coconut-flavoured milk or soya milk

Makes a 20cm/8in cake

Cook's tips
• Use a good-quality chocolate for the best results.
• You can buy dark (bittersweet) chocolate flavoured with coconut or orange, which would work particularly well in this cake.

1 Preheat the oven to 180°C/350°F/Gas 4. Grease and base-line a deep 20cm/8in round cake tin (pan). Melt the chocolate in a heatproof bowl over a pan of hot water.

2 Put the prunes in a food processor and process until roughly chopped. Add a little of the egg and process again. Continue until the mixture is light in colour and smooth, adding a little more of the egg if necessary to blend the mixture. With the machine still running, add the remaining beaten egg.

3 Scrape the mixture into a bowl. Gradually beat in the coconut oil, then fold in the chocolate. Sift together the flours and baking powder, then fold into the batter, alternating with the milk.

4 Spoon the mixture into the cake tin, level the surface with a spoon, then bake for 35–40 minutes, or until the cake is firm to the touch. Leave to cool on a wire rack.

NUTRITIONAL INFORMATION: Energy 4666kcal/19782kJ; Protein 101.8g; Carbohydrate 922.2g, of which sugars 816.4g; Fat 89.2g, of which saturates 41.7g; Cholesterol 717mg; Calcium 1579mg; Fibre 164.5g; Sodium 1074mg.

SEMOLINA AND COCONUT CAKE

Deceptively easy to make, this no-bake cake uses coconut water and takes no more than 20 minutes to prepare. It is both dairy- and gluten-free.

250g/9oz/1¼ cups caster (superfine) sugar
475ml/¾ pint/2 cups coconut water
½ cinnamon stick
120ml/4fl oz/½ cup coconut oil
175g/6oz/1 cup coarse semolina
25g/1oz/¼ cup blanched almonds
15ml/1 tbsp pine nuts
2.5ml/½ tsp ground cinnamon

Serves 6–8

Variation
For a larger cake to serve 8, simply double the quantity of ingredients and spoon the mixture into a 20–3cm/8–9in round cake tin (pan).

1 Put the sugar in a heavy pan, pour in the water and add the cinnamon stick. Bring to the boil, stirring until the sugar dissolves, then boil without stirring for about 4 minutes to make a syrup. Remove and discard the cinnamon stick.

2 Meanwhile, heat the oil in a separate heavy pan. When it is hot, add the semolina and stir until it turns light brown. Lower the heat, add the almonds and pine nuts and brown for 2–3 minutes, stirring continuously. Take off the heat and set aside.

3 Wearing an oven mitt, gradually add the hot syrup to the semolina mixture, stirring continuously. It will probably spit at this point, so stand well away. Return to a gentle heat and stir until the syrup has been absorbed and the mixture looks smooth.

4 Remove from the heat, cover it with a clean dish towel and let it stand for 10 minutes. Scrape the mixture into a 15–18cm/6–7in round cake tin (pan). When cold, unmould it on to a plate and dust it with the cinnamon.

NUTRITIONAL INFORMATION: Energy 326kcal/1373kJ; Protein 4.6g; Carbohydrate 49.9g, of which sugars 36.4g; Fat 13.4g, of which saturates 8.9g; Cholesterol 0mg; Calcium 21mg; Fibre 2.4g; Sodium 154mg.

LAVENDER COCONUT WHOOPIE PIES

Pretty and purple, these whoopie pies are a delight for the eye as well as the tastebuds. The coconut milk lends itself perfectly as a carrier for the lavender oils.

FOR THE CAKES
115g/4oz/8 tbsp coconut oil, soft but not liquid
150g/5oz/generous ½ cup coconut sugar or soft light brown sugar
50g/2oz/¼ cup caster (superfine) sugar
seeds of 1 vanilla pod (bean)
2 eggs
350g/12oz/3 cups plain (all-purpose) flour
7.5ml/1½ tsp bicarbonate of soda (baking soda)
a pinch of salt
150ml/¼ pint/⅔ cup coconut milk
3–4 drops edible lavender oil

FOR THE FILLING
30ml/2 tbsp lavender honey
475ml/16fl oz/2 cups whipped coconut cream (see page 26)

FOR THE ICING AND DECORATION
150g/5oz/1¼ cups icing (confectioners') sugar
25ml/1½ tbsp coconut water or cold water
5ml/1 tsp lavender food colouring
1–2 drops lavender extract
20g/¾oz fresh lavender flowers

Makes 12

Cook's tip
You can make lavender oil at home by putting edible lavender flowers in a bottle of light olive oil and leaving it to infuse for 1 month.

1 Preheat the oven to 180°C/350°F/Gas 4. Line two baking trays with baking parchment or silicone mats.

2 For the cakes, mix together the coconut oil, sugars and vanilla seeds. Whisk in the eggs, one at a time.

3 In a separate bowl, sift the flour with the bicarbonate of soda and salt. In a third bowl, stir the coconut milk and lavender oil together. Fold half of the dry ingredients into the sugar mixture. Mix in the coconut milk mixture, then the remaining dry ingredients.

4 Using a piping (pastry) bag fitted with a large plain nozzle, pipe 24 5cm/2in rounds of batter in total, 12 on each baking tray, spacing them well apart. Bake for 10–12 minutes, or until the cakes bounce back when gently pressed. Transfer to a wire rack to cool.

5 For the filling, fold the honey into the whipped coconut cream. Mix the icing ingredients together to form a smooth paste. Using a piping bag fitted with a star-shaped nozzle, pipe some filling on to the flat side of one cake and top with the flat side of another. Repeat to make 12 whoopie pies. Spread icing over the tops and decorate with lavender flowers.

NUTRITIONAL INFORMATION: Energy 303kcal/1282kJ; Protein 3.2g; Carbohydrate 60.6g, of which sugars 38.3g; Fat 7g, of which saturates 5.7g; Cholesterol 0mg; Calcium 82mg; Fibre 1.2g; Sodium 314mg.

LOW-FAT ORANGE AND COCONUT OATIES

These coconutty cookies are so delicious that it is difficult to believe that they are healthy too. Flavoursome and wonderfully crunchy, the whole family will love them.

175g/6oz/¾ cup clear honey
120ml/4fl oz/½ cup coconut water
90g/3½oz/1 cup rolled oats, lightly toasted
115g/4oz/1 cup plain (all-purpose) flour
50g/2oz/¼ cup coconut sugar
50g/2oz/¼ cup caster (superfine) sugar
finely grated rind of 1 orange
5ml/1 tsp bicarbonate of soda (baking soda)

Makes about 16

Variation
For an extra coconut hit, you could replace 15ml/ 1 tbsp of the rolled oats with an equal quantity of desiccated (dry unsweetened shredded) coconut.

1 Preheat the oven to 180°C/350°F/ Gas 4. Line two baking sheets with baking parchment.

2 Put the honey and coconut water in a small pan and simmer over a low heat for 8–10 minutes, stirring occasionally, until the mixture is thick and syrupy.

3 Put the oats, flour, sugars and orange rind into a bowl. Mix the bicarbonate of soda with 15ml/1 tbsp boiling water and add to the flour mixture, together with the honey and orange syrup. Mix well with a wooden spoon.

4 Place spoonfuls of the mixture on to the prepared baking sheets, spaced slightly apart, and bake for 10–12 minutes, or until golden brown. Leave to cool on the sheets for 5 minutes before transferring to a wire rack to cool completely.

NUTRITIONAL INFORMATION: Energy 104kcal/442kJ; Protein 1.6g; Carbohydrate 24.6g, of which sugars 15.4g; Fat 0.6g, of which saturates 0g; Cholesterol 0mg; Calcium 16mg; Fibre 1g; Sodium 108mg.

COCONUT FRUIT SLICE

A double-layered cookie for which the topping is a combination of dried fruit and coconut oil mixed with grated carrot to keep it moist.

90g/3½oz/7 tbsp coconut oil
75g/3oz/scant ½ cup caster (superfine) sugar
1 egg yolk
115g/4oz/1 cup plain (all-purpose) flour
30ml/2 tbsp self-raising (self-rising) flour
30ml/2 tbsp desiccated (dry unsweetened shredded) coconut
icing (confectioners') sugar, for dusting

FOR THE TOPPING
30ml/2 tbsp ready-to-eat pitted prunes, chopped
30ml/2 tbsp sultanas (golden raisins)
50g/2oz/½ cup ready-to-eat dried pears, chopped
25g/1oz/¼ cup walnuts, chopped
75g/3oz/⅔ cup self-raising (self-rising) flour
5ml/1 tsp ground cinnamon
2.5ml/½ tsp ground ginger
2.5ml/½ tsp bicarbonate of soda (baking soda)
90g/3½oz/scant ½ cup coconut sugar
175g/6oz/generous 1 cup grated carrots
1 egg, beaten
75ml/5 tbsp coconut oil

Makes 12–16

1 Preheat the oven to 180°C/350°F/Gas 4. Line a 28 x 18cm/ 11 x 7in shallow baking tin (pan) with baking parchment. In a large mixing bowl beat together the coconut oil, sugar and egg yolk until smooth and creamy.

2 Stir in the flours and shredded coconut and mix together well. Press into the base of the prepared tin, using your fingers to spread out the dough evenly. Bake for 15 minutes, or until firm and light brown.

3 To make the topping, mix together all the ingredients and spread over the cooked base. Bake for about 35 minutes, or until firm. Cool completely in the tin before cutting into bars or squares. Dust with icing sugar.

NUTRITIONAL INFORMATION: Energy 217kcal/907kJ; Protein 2.9g; Carbohydrate 25.4g, of which sugars 15g; Fat 12.2g, of which saturates 8.7g; Cholesterol 52mg; Calcium 51mg; Fibre 1.8g; Sodium 36mg.

MINI LAMINGTONS

These delicious individual cakes are a favourite treat in Australia. The key to success is the light and moist coconut sponge. Here they are covered in dark chocolate icing.

coconut oil, for greasing
3 eggs
100g/3¾oz/generous ½ cup caster (superfine) sugar
100g/3¾oz/scant 1 cup self-raising (self-rising) flour
35g/1¼oz/⅓ cup cornflour (cornstarch)
15ml/1 tbsp coconut oil, melted
45ml/3 tbsp coconut water
300g/11oz/4 cups desiccated (dry unsweetened shredded) coconut

FOR THE ICING
15ml/1 tbsp coconut oil
375g/13oz/3¼ cups icing (confectioners') sugar, sifted
150g/5oz dark (bittersweet) chocolate (55% cocoa solids), chopped
90ml/6tbsp coconut-flavoured milk or milk

Makes 12–15

1 Preheat the oven to 160°C/325°F/Gas 3. Grease a 20 x 30cm/ 8 x 12in cake tin (pan) and line it with baking parchment.

2 For the cake, beat the eggs until they begin to froth. Add the sugar and beat until light and fluffy. Sift in the flour and cornflour and fold in with a metal spoon. Put the coconut oil and coconut water in a small pan and heat until melted and warm. Fold into the cake mixture. Pour into the tin and bake for 25–30 minutes, until just set and coming away slightly from the edge of the tin.

3 Allow to cool in the tin for 10 minutes before turning out the cake on to a wire rack to cool completely. Cut the cake into 12–15 pieces. Put the coconut in a wide, shallow dish.

4 Place all of the icing ingredients in a heatproof bowl set over a pan of simmering water. Whisk constantly to make an emulsified icing. Turn off the heat, leaving the bowl over the hot water.

5 Use a dipping fork to dip each cake into the icing, wiping off the excess along the edge of the bowl. Gently turn in the coconut to coat. Place on a wire rack and leave to set for 15 minutes.

NUTRITIONAL INFORMATION: Energy 362kcal/1517kJ; Protein 4.1g; Carbohydrate 48.3g, of which sugars 41g; Fat 18.2g, of which saturates 14.1g; Cholesterol 47mg; Calcium 41mg; Fibre 4.3g; Sodium 29mg.

COCONUT DATE ROLLS

Incredibly simple to make, these contain just two healthy ingredients, yet make delicious and attractive after-dinner treats to enjoy with coffee.

36 Medjool dates
15ml/1 tbsp coconut water
150g/5oz/2 cups desiccated (dry unsweetened shredded) coconut

Makes 36

NUTRITIONAL INFORMATION:
Energy 80kcal/336kJ; Protein 0.9g; Carbohydrate 14g, of which sugars 14g; Fat 2.6g, of which saturates 2.3g; Cholesterol 0mg; Calcium 10mg; Fibre 1.9g; Sodium 3mg.

1 Remove the skins from the dates. Cut them in half and remove the stones (pits).

2 Place the dates in a small pan with the coconut water. Simmer for 5 minutes, or until softened. Push the dates through a sieve (strainer) using the back of a spoon. Roll the pulped dates into small balls, about the same size as the dates were originally.

3 Place the coconut in a shallow bowl and roll the date balls in it. Place in paper cups and serve, or store in an airtight container.

COCONUT SWEETS

These chewy coconut treats are a favourite with children. They can be made plain but the lime juice gives them an interesting twist that adults cannot resist either.

50g/2oz/²⁄₃ cup desiccated (dry unsweetened shredded) coconut
105ml/7 tbsp coconut water
175g/6oz/¾ cup coconut sugar or light muscovado (brown) sugar
juice of ½ lime

Makes 25

NUTRITIONAL INFORMATION:
Energy 40kcal/169kJ; Protein 0.2g; Carbohydrate 7.4g, of which sugars 7.7g; Fat 1.2g, of which saturates 1.1g; Cholesterol 0mg; Calcium 2mg; Fibre 0.5g; Sodium 11mg.

1 Line a tray with baking parchment. Put the shredded coconut in a pan with the coconut water and sugar. Heat until the sugar dissolves.

2 Stir in the lime juice and increase the heat. Cook, stirring with a wooden spoon, until the mixture has thickened and become dark golden brown.

3 Drop spoonfuls of the mixture on to the lined tray, pressing the mixture down with the back of the spoon to flatten it lightly into chunky, irregular pieces. Leave to cool before eating.

INDEX

Apricot and Coconut Dessert 94
arrack 21
avocados: Chilled Avocado Soup with Coconut 50
Coconut Guacamole 58

Barley Coconut Risotto with Squash and Leeks 68
Batida de Côco 21
beauty 36–7
beef: Beef, Mushroom and Coconut Salad 65
Madras Beef Curry with Coconut Rice 88
black beans: Spicy Black Bean Coconut Burgers 67
breads: Barbadian Coconut Sweet Bread 102
Quinoa and Coconut Curry with Seeded Flatbread 70
Sultana, Coconut and Walnut Bread 103

cakes 104–6, 110
carrots: Coconut and Pineapple Carrot Cake 104
Coconut Fruit Slice 109
chicken: Adobo Chicken and Pork with Coconut 86
Chicken Breasts with Ham and Coconut 79
Chicken Rice Soup with Lemon Grass and Coconut 55
Chicken, Split Pea and Coconut Koresh 80
Citrus and Coconut Chicken Coleslaw 64
Coconut Chicken Biryani 81
chickpeas: Chickpea Coconut Pilau 69
Tofu Falafels with Coconut 61
Chocolate and Banana Fool 92
Chocolate, Coconut and Prune Cake 105
coconut 6–7
creamed 17, 76
desiccated 14, 15, 45, 60, 102, 108, 109, 111
flaked 14, 15, 57
mature 7, 12–13, 15, 17, 101
shredded 14, 15, 23, 60, 72, 102, 109, 111
young 7, 8, 9, 12, 98, 101
coconut alcohol 21
coconut aminos 20, 21, 63, 66, 72, 77, 86
coconut feeder 39
coconut butter 14–15, 49
coconut cream 16–17, 42, 52, 58, 63, 64, 76, 81, 84, 90, 95
coconut cream, whipped 26, 92, 94, 100, 104, 107
coconut dulce de leche 27
coconut flour 17, 18, 19, 31, 35, 42, 47, 61, 67, 102, 105
coconut frosting 27
Coconut Fruit Slice 109
coconut leaves 38, 39
coconut mayonnaise 26, 74
coconut milk 8, 16, 17, 18, 20, 21, 22, 26, 27, 54, 57, 63, 76, 90, 92, 96, 97, 107

coconut oil 9–11, 14, 15, 23, 24–7, 29, 30, 31, 32, 33, 34, 35, 36, 37, 41, 45, 46, 47, 48, 49, 52, 54, 57, 58, 59, 60, 61, 64, 65, 66, 67, 68, 69, 70, 73, 74, 75, 76, 77, 78, 79, 80, 81, 82, 83, 84, 86, 87, 88, 92, 99, 100, 101, 102, 103, 105, 106, 107, 109, 110
coconut pastry 25
Coconut Pudding, Young 98
coconut roots 39
coconut salad dressing 25
coconut shells 38, 39
coconut sugar 18–19, 24, 27, 31, 35, 42, 43, 45, 46, 47, 52, 58, 62, 63, 65, 66, 72, 77, 83, 86, 90, 92, 94, 95, 96, 97, 99, 101, 102, 104, 107, 108, 109, 111
coconut syrup 19, 45, 47, 103
coconut trunks 38, 39
coconut vinegar 20, 24, 25, 53, 58, 59, 63, 66, 72, 73, 86
coconut water 7, 8–11, 12, 13, 14, 20, 21, 22–3, 28–9, 31, 34, 41, 42, 43, 45, 46, 47, 50, 51, 52, 53, 54, 55, 56, 57, 58, 59, 60, 62, 67, 68, 69, 70, 73, 74, 77, 78, 80, 81, 84, 86, 88, 90, 91, 94, 97, 98, 99, 100, 101, 103, 106, 107, 108, 110, 111
coconut-flavoured milks and drinks 20, 43, 45, 46, 95, 103, 105, 110
coir 38
Coleslaw, Citrus and Coconut Chicken 64
cookies 108, 109
copra 9
crab: Crab and Tofu Stir-fry with Coconut 77
Crab in Coconut Vinegar 72
curries 70, 88
Custard, Coconut 96

dates: Coconut Date Rolls 111
Coconut Porridge with Dates and Nuts 43
desserts 90–101
diabetes 31
diets 31, 34
Dip, Coconut and Herb Chilli 58
drinks: Batida de Côco 21
Coconut and Hazelnut Smoothie 42
Coconut and Passion Fruit Ice 42
duck: Stir-fried Duck with Noodles and Coconut 82

eggs: Mixed Vegetable Coconut Omelette 48
Coconut Scrambled Eggs 49
electrolytes 7, 8, 28–9

fish: Baked Coconut Monkfish with Potatoes 74
Coconut Ceviche with Tomato Salsa 73

Coconut Hake with Lemon and Chilli 75
Fish in Spiced Coconut Milk 76
Fragrant Thai Fish and Coconut Soup 53
Marinated Sashimi-style Coconut Tuna 72
Salmon Ceviche with Coconut 23

Gado Gado Salad with Coconut 63
Granola, Coconut 45
Guacamole, Coconut 58

health benefits 30–1
heart health 32–3
hydration 28–9

Ice, Watermelon and Coconut 91
Ice Cream, Coconut and Lemon Grass 90

lamb: Indian Lamb Soup with Rice and Coconut 57
Spiced Coconut Lamb with Tomatoes and Peppers 87
Lamingtons, Mini 110
lemon grass: Chicken Rice Soup with Lemon Grass and Coconut 55
Coconut and Lemon Grass Ice Cream 90
lemons: Coconut and Lemon Risotto 23
Coconut Hake with Lemon and Chilli 75
Coconut, Lemon and Raisin Pancakes 47

mango: Coconut Mango Stacks with Raspberry Coulis 100
Muesli, Luxury Coconut 45
Muffins, Crunchy Coconut Breakfast 46
mushrooms: Beef, Mushroom and Coconut Salad 65

nata de coco 20, 21
noodles: Stir-fried Duck with Noodles and Coconut 82

oats: Coconut Porridge with Dates and Nuts 43
Low-fat Orange and Coconut Oaties 108
Mixed Berry and Coconut Quinoa Porridge 43
oil pulling 37
oranges: Citrus and Coconut Chicken Coleslaw 64
Low-fat Orange and Coconut Oaties 108

Paella, Spicy Coconut 78
Pancakes, Coconut, Lemon and Raisin 47
Patties, Coconut 60
pineapple: Coconut and Pineapple Carrot Cake 104
Pineapple with Papaya and Coconut Sauce 99

plums: Caramelized Plums with Sticky Coconut Rice 95
polenta: Griddled Coconut Polenta with Tangy Pebre 59
pork: Adobo Chicken and Pork with Coconut 86
Stuffed Pork with Coconut 84
Sweet and Sour Coconut Pork with Vegetables 83
Tamarind Pork, Vegetable and Coconut Soup 56
porridge: Coconut Porridge with Dates and Nuts 43
Mixed Berry and Coconut Quinoa Porridge 43
prawns: Coconut and Seafood Soup 54
prunes: Chocolate, Coconut and Prune Cake 105
Coconut Fruit Slice 109
Pumpkin Soup with Coconut 52

quinoa: Mixed Berry and Coconut Quinoa Porridge 43
Quinoa and Coconut Curry with Seeded Flatbread 70

rice: Caramelized Plums with Sticky Coconut Rice 95
Chicken Rice Soup with Lemon Grass and Coconut 55
Chickpea Coconut Pilau 69
Coconut Fruit Slice 109
Coconut Chicken Biryani 81
Indian Lamb Soup with Rice and Coconut 57
Madras Beef Curry with Coconut Rice 88
Spicy Coconut Paella 78
Thai Coconut Rice Salad 62
RNI (Reference Nutrient Intake) 29

salads 62, 63, 65
Semolina and Coconut Cake 106
soups 41, 50–7
Coconut and Seafood Soup 54
Summer Vegetable and Coconut Soup 51
Stir-fry, Vegetable with Coconut 66
Sweets, Coconut 111

Tamarind Pork, Vegetable and Coconut Soup 56
Tapioca Pudding, Coconut 97
Tarts, Coconut 101
tofu: Crab and Tofu Stir-fry with Coconut 77
Tofu Falafels with Coconut 61
tomatoes: Coconut Ceviche with Tomato Salsa 73
Spiced Coconut Lamb with Tomatoes and Peppers 87
Trifle, Coconut and Coffee 92

Watermelon and Coconut Ice 91
weight loss 34–5
Whoopie Pies, Lavender Coconut 107

Young Coconut Pudding 98